Synthesis Lectures on Digital Circuits & Systems

Series Editor

Mitchell A. Thornton, Southern Methodist University, Dallas, USA

This series includes titles of interest to students, professionals, and researchers in the area of design and analysis of digital circuits and systems. Each Lecture is self-contained and focuses on the background information required to understand the subject matter and practical case studies that illustrate applications. The format of a Lecture is structured such that each will be devoted to a specific topic in digital circuits and systems rather than a larger overview of several topics such as that found in a comprehensive handbook. The Lectures cover both well-established areas as well as newly developed or emerging material in digital circuits and systems design and analysis.

Steven F. Barrett

Arduino VII

Industrial Control

Springer

Steven F. Barrett
University of Wyoming
Laramie, WY, USA

ISSN 1932-3166 ISSN 1932-3174 (electronic)
Synthesis Lectures on Digital Circuits & Systems
ISBN 978-3-031-68608-5 ISBN 978-3-031-68609-2 (eBook)
https://doi.org/10.1007/978-3-031-68609-2

This Springer imprint is published by the registered company Springer Nature Switzerland AG
The registered company address is: Gewerbestrasse 11, 6330 Cham, Switzerland

Preface

This book is about the Arduino microcontroller and the Arduino concept. The visionary Arduino team of Massimo Banzi, David Cuartielles, Tom Igoe, Gianluca Martino, and David Mellis launched a new innovation in microcontroller hardware in 2005, the concept of open-source hardware. Their approach was to openly share the details of microcontroller-based hardware design platforms to stimulate the sharing of ideas and promote innovation. This concept has been popular in the software world for many years. Their efforts resulted in a global phenomenon of making computing accessible for all.

I was quite excited when I heard Arduino was extending the concept of accessible computing to the industrial and Internet of Things (IoT) sectors. Originally I planned a book including both of the Opta series of programmable logic controllers (PLCs) and the Portenta Machine Controller (PMC). As the book evolved it became quite clear there was too much information for a single text. Instead, a complementary set of books was planned: *Arduino VII: Industrial Control* and *Arduino VIII: Machine Control*. Although the books are a complementary set, each is independent in the information contained.

This book, *Arduino VII: Industrial Control*, is an accessible primer on industrial control and programmable logic controller concepts for those without a deep instrumentation background. An understanding of basic circuit theory is an appropriate prerequisite for the book. The three main goals of the book are: explore accessible Arduino Opta industrial control products; learn the fundamentals of programming using ladder logic; and explore related sensors and interface concepts. We use multiple examples throughout the book and conclude with an instrumented greenhouse project. Throughout the book we concentrate on remote, direct current (DC) powered systems. We develop systems that operate on positive polarity (e.g. supplied by solar panels and batteries).

Approach of the Book

The book has been divided into a series of six chapters to accomplish the book's goals. The book follows these chapters:

- Chapter 1. Operational Technology and the Arduino Opta
- Chapter 2. Opta Features
- Chapter 3. Arduino PLC IDE and Ladder Logic
- Chapter 4. Input Sensors, Output Actuators, and Interfacing
- Chapter 5. Application: IoT Greenhouse
- Chapter 6. Opta Expansions[1]
- Appendix A. Safety
- Appendix B. Embedded Systems Design

Throughout the book, we provide numerous hardware and software examples. A tutorial on safety concepts is readily available in Appendix A and referenced throughout the book. We recommend reading this appendix first (now) and regularly as you progress through the book. Appendix B provides a tutorial on system design concepts and tools.

For completeness and independence, this volume contains tutorial information contained in some of the other volumes in the Arduino series and related works completed for Morgan and Claypool and Springer Nature. Chapter footnotes identify the source of this information contained elsewhere in the series. The book series thus far includes:

- Arduino I: Getting Started
- Arduino II: Systems
- Arduino III: Internet of Things
- Arduino IV: DIY Robots–3D Printing, Instrumentation, Control
- Arduino V: AI and Machine Learning
- Arduino VI: Bioinstrumentation
- Arduino VII: Industrial Control

In the rapidly evolving Arduino world, I anticipate other books in the series. As mentioned, *Arduino VIII: Machine Control* is in development.

Laramie, WY, USA
April 2025

Steven F. Barrett

[1] I am thankful to Chuck Glaser, Editor, and Boopalan Renu, Production Editor, who paused the production schedule to allow inclusion of this chapter.

Acknowledgments

A number of people have made this book series possible. I would like to thank Massimo Banzi of the Arduino design team for his support and encouragement in writing the first edition of this book: *Arduino Microcontroller: Processing for Everyone!*.

I would also like to acknowledge Joel Claypool for his publishing expertise and support on a number of writing projects. His vision and expertise in the publishing world have made this book possible. Joel "retired" in September 2022 after 40+ years of service to the U.S. Navy and the publishing world. On behalf of the multitude of writers you have provided a chance to become published authors, we thank you!

I would also like to thank Charles (Chuck) Glaser, Editorial Director at Springer Nature, for his encouragement and support on this project. If you have a good idea for a book, I highly recommend contacting Chuck. He will assist you in converting your idea into a finished, professional book product.

I would also like to thank Boopalan Renu of Total Service Books Production for his expertise in converting the final draft into a finished product. You provide outstanding service.

Finally, as most importantly, I would like to thank my wife and best friend of many (almost 50) years, Cindy.

Laramie, WY, USA
April 2025

Steven F. Barrett

Contents

About the Author

Steven F. Barrett Ph.D., P.E., received the BS Electronic Engineering Technology from the University of Nebraska at Omaha in 1979, the M.E.E.E. from the University of Idaho at Moscow in 1986, and the Ph.D. from The University of Texas at Austin in 1993. He was formally an active duty faculty member at the United States Air Force Academy, Colorado and is now the Associate Dean for Undergraduate Programs at the University of Wyoming and Professor of Electrical and Computer Engineering. He is a member of IEEE (Life Senior) and Tau Beta Pi (chief faculty advisor). His research interests include digital and analog image processing, computer-assisted laser surgery, and embedded controller systems. He is a registered Professional Engineer in Wyoming and Colorado. He co-wrote with Dr. Daniel Pack several textbooks on microcontrollers and embedded systems. In 2004, Barrett was named "Wyoming Professor of the Year" by the Carnegie Foundation for the Advancement of Teaching and in 2008 was the recipient of the National Society of Professional Engineers (NSPE) Professional Engineers in Higher Education, Engineering Education Excellence Award. In 2023, Barrett received the National Council of Examiners for Engineering and Surveying (NCEES) Distinguished Examination Service Award.

Operational Technology and the Arduino Opta 1

Objectives: After reading this chapter, the reader should be able to do the following:

- Define Information Technology (IT) and Operational Technology (OT);
- Describe the features of a programmable logic controller (PLC) based OT system;
- Provide a working definition of the Industrial Internet of Things (IIoT);
- Describe different variants of the Arduino Opta micro PLC; and
- Construct a portable lab environment for the Arduino Opta micro PLC.

1.1 Overview

In this chapter we begin our exploration of the Operational Technology (OT) world. We start with a basic introduction to the Internet of Things (IoT). Within IoT there is a close relationship between Information Technology (IT) and Operational Technology (OT). We explore this relationship in some detail. The reader is assumed to have a solid grounding in basic IT concepts.[1] The pervasiveness of IoT is then examined in industry or the Industrial Internet of Things (IIoT). We then shift our focus to OT and basic PLC concepts. We conclude the chapter with an introduction to the Arduino Opta micro PLC.

[1] A basic introduction to IT concepts is provided in "Arduino III: Internet of Things," S.F. Barrett, Springer Nature, 2021. Portions of this chapter have been adapted with permission.

S. F. Barrett, *Arduino VII*, Synthesis Lectures on Digital Circuits & Systems,
https://doi.org/10.1007/978-3-031-68609-2_1

1.2 Internet of Things–IoT

The term Internet of Things was first used by Kevin Ashton in a 1999 Proctor and Gamble presentation. Mr. Ashton's presentation discussed concepts on using the existing internet infrastructure to support P&G's supply chain (Greer, Hanes). From this early start, applications within business and industry have become quite pervasive. A review of the literature provides a feature list describing the Internet of Things systems concept (Rajkumar, Hanes, Greer):

- An IoT system connects things to the internet;
- Each thing or device has its own unique identifier or address;
- Communication between things is provided via the internet;
- An IoT system provides for interrelated and integrated computing devices and physical processes;
- An IoT system provides the ability to measure, process, and transfer to and from remote locations; and
- IoT processes are monitored, coordinated, and controlled.

Interestingly the concept of Cyber–physical Systems (CPS) share many of the same features. The National Institute of Standards and Technology (NIST) performed a study to examine the relationship between IoT and CPS and noted although the concepts originated in different industries, they are substantially equivalent concepts. A unified perspective of the two concepts was provided: "Internet of Things and Cyber–Physical Systems comprise interacting logical, physical, transducer, and human components engineered for function through integrated logic and physics (Greer)."

1.3 Information Technology Versus Operational Technology

A key concept within IoT is the close relationship between IT and OT. The relationship between IT and OT are shown in Fig. 1.1.

IT communications usually consist of short, frequent communications that are broken into packets and communicated globally. IT provides a wide variety of message traffic including e-mails, requests and response for information from websites, and multiple other types. IT technology developments are rapidly evolving with vulnerabilities well known, documented, combatted, and corrected (Hanes).

Operational Technology (OT) provides for process control within many areas of industry. As shown in Fig. 1.1, industrial safety and security are intertwined. OT communications are typically short, point–to–point communications on a factory floor or within an industrial process. Monitoring via a Supervisory Control and a Data Acquisition (SCADA) system is typically performed within a local and confined zone. Although OT developments are

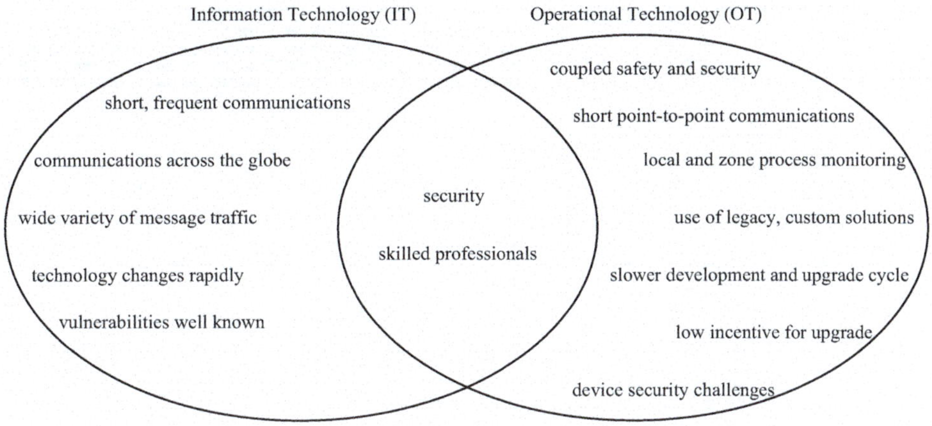

Fig. 1.1 Information Technology (IT) versus Operational Technology (OT) (Hanes)

actively taking place, adoption timelines are slower than with IT. Since OT governs a proprietary and custom solution for a given industrial process, typically there is a low incentive for technology upgrade. Although many IT and OT concepts are related but different, IT and OT both enjoy the dedication of skilled professional practitioners (Hanes).

IT and OT share the requirement for robust security protection and countermeasures. Many of the security concepts discussed for IT also apply for OT. In the industrial world IT and OT systems are often linked to share information among related processes. For example, a remote oil drilling platform may be controlled via OT processes. If an oil company has multiple remote platforms, they may be linked via IT processes to share production data. Some form of isolation, an "air gap," is typically provided between IT and OT related processes for security purposes. This helps prevent a nefarious actor from accessing a critical industrial process via the internet (Hanes).

1.4 Operational Technology

Operational Technology is used to control industrial processes. The fundamental OT building block is the programmable logic controller (PLC). A PLC diagram is provided in Fig. 1.2a.

A PLC is an industrial hardened microcontroller. As shown in Fig. 1.2a, a PLC is typically a rack mounted collection of modules. Each module provides a critical subsystem for the PLC. The PLC subsystems share many of the same functions typically found in most microcontrollers. For example, a typical PLC system consists of power supply, central processing unit (CPU), serial communications, analog input, digital input and output, and timer modules. A custom system is assembled by choosing modules to meet system requirements.

PLC systems are typically programmed using ladder logic techniques. A ladder logic program resembles a ladder with two vertical side rails linked by a number of rungs. As

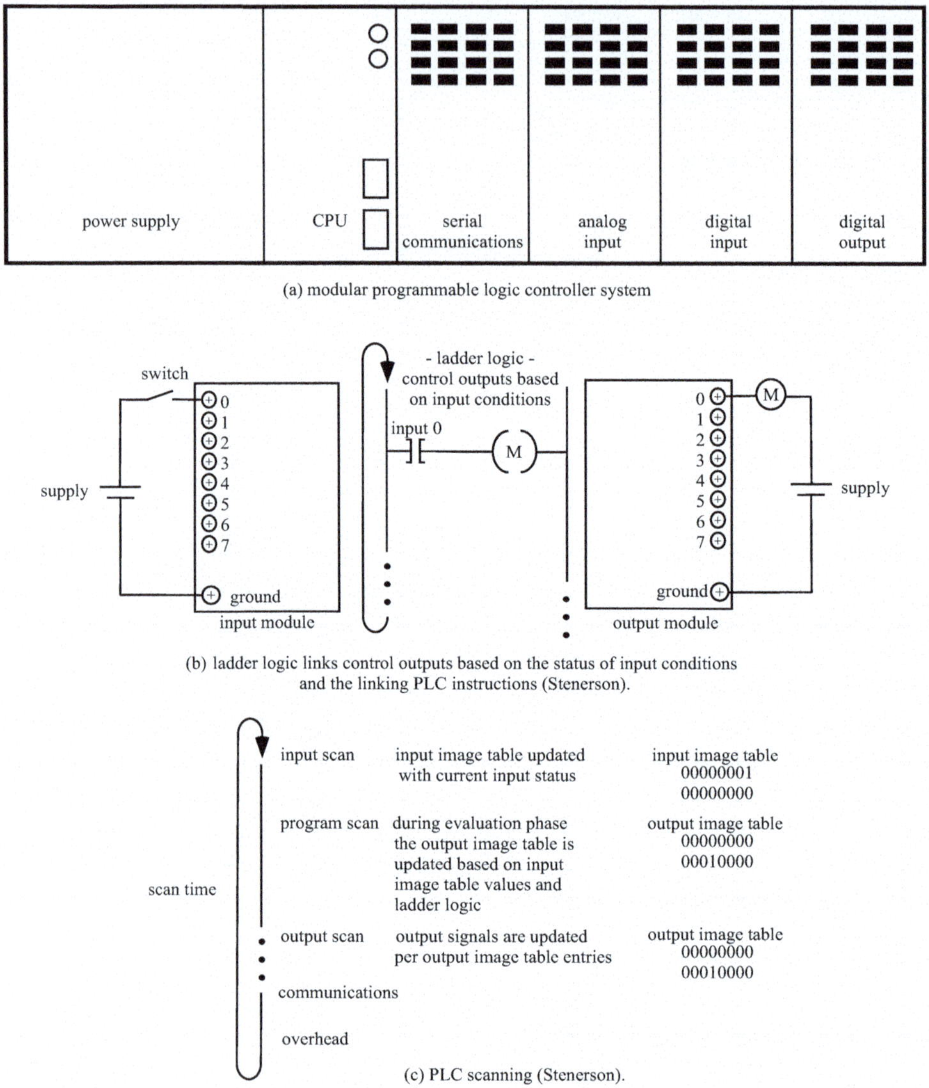

Fig. 1.2 PLC overview (Stenerson)

shown in Fig. 1.2b, the real world inputs (switches, sensors, etc.) are interfaced to the PLC via input modules. Output real world devices such as indicators, audible alarms, motors, actuators, etc. are interfaced to the PLC via output modules. The PLC ladder logic rungs represent steps of a program that link input device conditions to output control signals.

As shown in Fig. 1.2c, a ladder logic program goes through a scan consisting of multiple stages. The scan begins with an input scan. During the input scan, the status of inputs is checked and an input image table is updated in the PLC CPU memory. The input status is fixed in the input image table for the remainder of the scan time. With input status updated, the program scan commences. This is called the evaluation phase where the output image table is updated based on the input image table values and the ladder logic rungs connecting input values to output control signals. Each rung in the ladder logic program is evaluated sequentially starting with the top rung and progressing down the ladder. With the completion of the evaluation stage, output signals are generated per the output image table. The final two steps of the scan include related serial communications and any required PLC housekeeping. Upon completion of the scan, the scan is repeated beginning again with the input scan (Stenerson).

1.5 IoT Architecture

The Internet of Things (IoT), as first described by Mr. Ashton in 1999, initiated the movement to provide a link between the IT and OT worlds. There are multiple models available to describe this vital link. Hanes et al. provides the model shown in Fig. 1.3a.

The model provides three layers linking IoT "things" to applications via internet–based communication channels. The "things" are the sensors and actuators interfacing to a physical world process. The sensors and actuators provide for the monitoring and control signals for the process. The application, which may be physically distant from the process, takes in as input the sensor information and provides output control signals based on the control algorithm (Hanes).

Example. I have always wanted to build a greenhouse. I find it quite fascinating that the sun's energy may be captured, stored, and employed at a later time to extend and stabilize the growing season for vegetables. Part of the fascination may be related to spending much of my life in northern climes (Newfoundland, Nebraska, North Dakota, Montana, and Wyoming).

Applying the IoT model described, the "things" of the greenhouse would be the sensors used to measure the vital signs of the greenhouse. For example, we might measure the following parameters: indoor temperature, outdoor temperature, humidity, soil moisture, stored water level, backup battery voltage level, etc. The actuator "things" of the greenhouse would be those devices used to change the greenhouse configuration: a vent fan when the indoor greenhouse temperature becomes too high, a water pump to mist the vegetables when appropriate conditions are met (e.g. plant soil too dry, etc.). An Arduino–based sketch may be developed to visualize and manage greenhouse properties. For example, the greenhouse indoor and outdoor temperatures may be logged and displayed over a long period of time (e.g. the winter months). The internet infrastructure with WiFi access may be used to allow the sending and receiving of greenhouse events. Also, a Bluetooth link to a cell phone might be helpful. In Chap. 5 we explore this IoT application in detail.

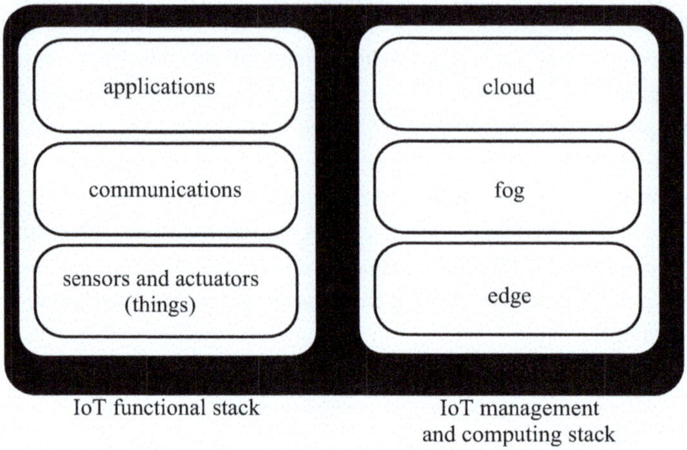

(a) Simplified IoT model (Hanes).

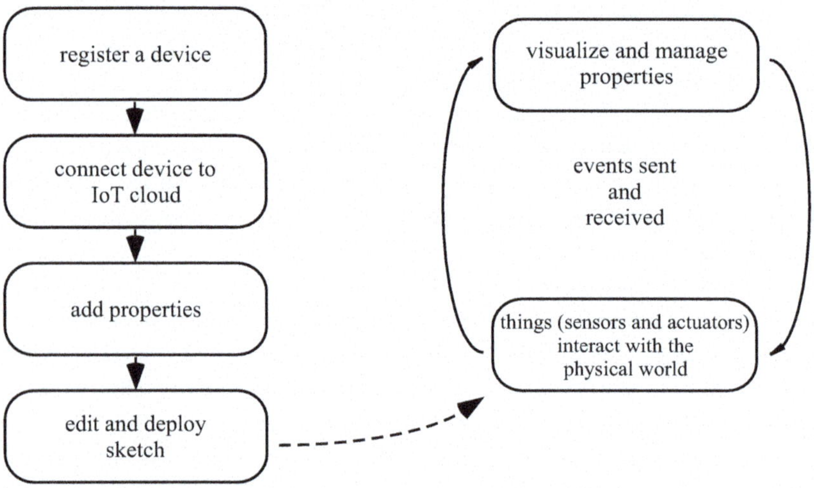

(b) Arduino IoT deployment model (arduino.cc).

Fig. 1.3 IoT models (Hanes, Arduino.cc)

1.6 IoT Technology

To support IoT deployment, a number of technologies have been developed to support project level IoT applications, smart home concepts, and industrial level Industrial Internet of Things (IIoT) applications. The dividing lines between these applications are blurry. It is more of a continuum of applications rather than categories of applications.

Smart Home Applications. A smart home uses technology to efficiently monitor and control home parameters such as temperature, humidity, lighting, security, lawn health, etc. In 2005 the Z–Wave Alliance was established to provide a standard configuration and control protocol for smart home applications. The Z–wave protocol provides for the wireless mesh networking of smart objects within a home. The protocol provides for a data rate between configured devices of 100 kbps. Devices communicate securely at frequencies of 908.4 or 916 MHz using AES 128 encryption.[2] Network activities are coordinated by a smart hub that is connected to the internet. The smart hub can control up to 232 devices within a home or small business environment at a range up to 328 ft. Each smart home network has a unique network identification and each device within the home has node identification. The node identification is provided using the IPv6 address space. This provides for non–interference between smart configured homes within a neighborhood (z-wavealliance.org).

1.7 Industrial Internet of Things (IIoT)

IoT technology has found its way into a number of industries as shown in Fig. 1.4. This merger of IoT concepts and processes applied to industry has resulted in the Industrial Internet of Things or IIoT. As an end of chapter assignment, we ask you to investigate one of these areas.

1.8 Cybersecurity

One of my favorite books is the *The Once and Future King* by T.H. White. It is the tale of the young boy, "the Wart," becoming King Arthur and the many adventures along the way. I have read this book every several years since I was young. Early in the book, White provides a description of the Wart's guardian's castle. He describes how the castle is protected from marauders by a moat (deep ditch) filled with water. To get access to the castle, a drawbridge is lowered across the moat and then raised again to secure the castle.[3]

There are many dangers surrounding a network or a computer on the internet or within an operational environment. As shown in Fig. 1.5 the dangers are in the form of malicious software (malware) or the nefarious efforts of computer hackers. These dangers and challenges include (Kurose, Levine, Lowe):

- botnet–network of infected computers controlled from an external source to perform coordinated nefarious activities on target computers;
- hackers–individuals who try to overcome computer protection measures and procedures to gain personal data;

[2] AES 128 is a data encryption standard.

[3] T.H. White, "The Once and Future King".

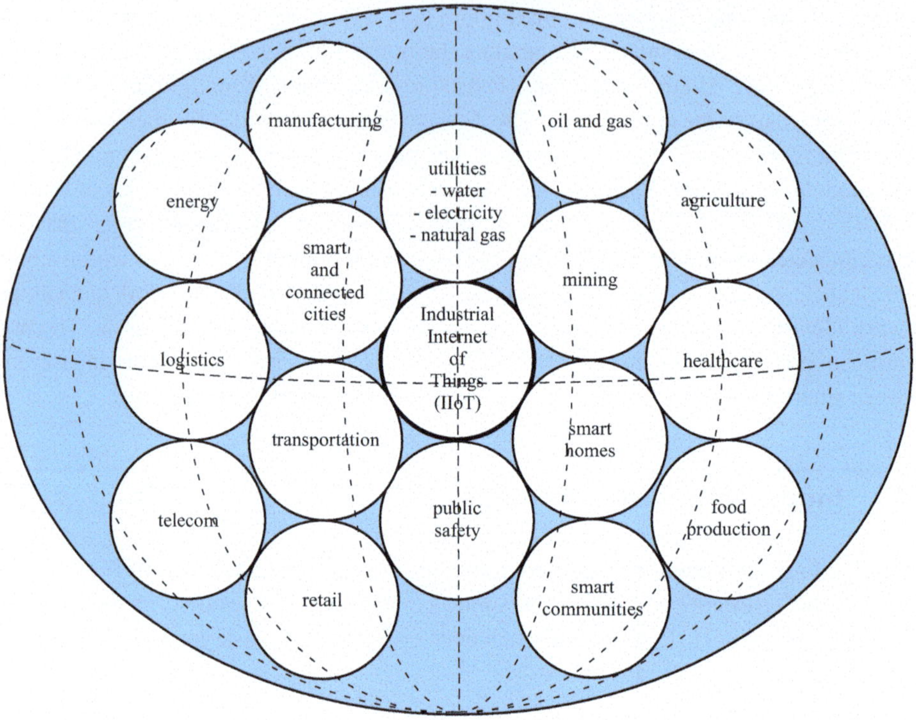

Fig. 1.4 IIoT applications

- ransomware–a computer attack where files are encrypted and held for ransom. If the ransom is paid, the files are returned to normal service;
- spyware–software that is accidently downloaded while browsing the internet. The software spies on your computer activities and reports back to its source;
- virus–a nefarious software program spread as an e-mail attachment. When the e-mail attachment is executed it goes to your computer's address bank and sends out e-mails with the virus program as an attachment masquerading as you. Using this technique, the virus may be spread to a number of computers. The nefarious intent of the virus may be activated by a specific event such as reaching a particular date and time; and
- worm–a worm creeps into a computer by means of flaws within network programs. Once onboard your computer, the worm looks for password and credit card information.

As in the analogy, the castle is protected from marauders by the surrounding moat and securing the drawbridge. The moat and drawbridge for a network and its computer assets include preventive countermeasures including (Kurose, Levine, Lowe):

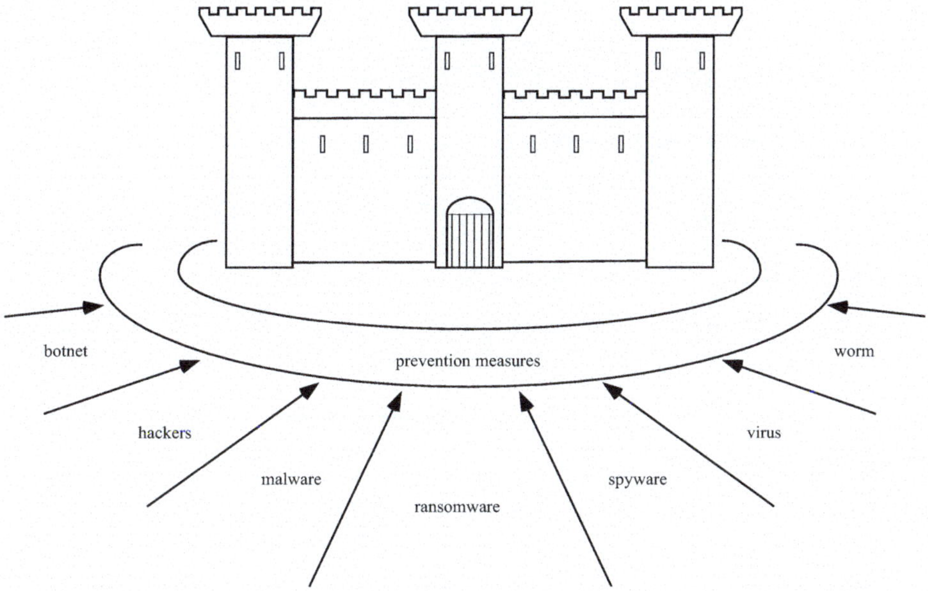

Fig. 1.5 Network threats

- Firewall–A firewall protects network resources from external dangers. It applies policies to determine message traffic that may enter a protected network.
- Antivirus programs (AVP)–Each computer on the network should have an AVP installed. The AVP should be current with all software updates applied.
- Operating system updates–Regular operating system updates are sent to computer users. These updates should be made when received. They may contain updates to correct a security flaw.
- Passwords–You should employ a strong password to protect your computer assets. IoT hardware devices are sometimes configured with a default password. You should replace the default password with a strong password.
- File backups–Computer files should be backed up on a regular basis.
- User awareness–Users should be skeptical of e-mails that appear questionable. An e-mail with an executable attachment should not be opened.

1.9 IoT and IIoT Security

IoT and IIoT security borrows many of the same measures from the IT world discussed earlier in the chapter. In addition, the International Society of Automation (ISA) and the International Electrotechnical Commission (IEC) have jointly developed a suite of security

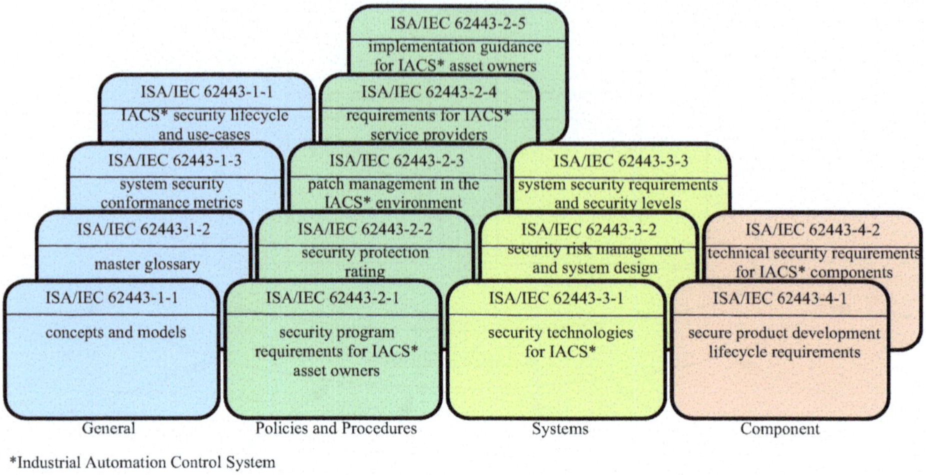

Fig. 1.6 ISA/IEC 62443 control system security (www.isa.org)

processes, procedures, and standards for control systems as shown in Fig. 1.6 (www.isa.org). The literature contains documentation of nefarious actors penetrating a secure system. Typically, these attacks have breached the "air gap" and vulnerabilities between the IT and OT components of the system.

1.10 Arduino Opta Micro Programmable Logic Controller

The Arduino company has partnered with the Finder company to develop and release a line of basic, yet powerful, Opta programmable logic controllers. The Opta PLCs consist of three variants as shown in Fig. 1.7.

Name	Opta Lite	Opta WiFi	Opta RS485
SKU	AFX00003	AFX00002	AFX00001
USB	USB-C	USB-C	USB-C
Ethernet	10/100BASE-T port	10/100BASE-T port	10/100BASE-T port
RS-485	n/a	half-duplex	half-duplex
Wi-Fi	n/a	802.11 b/g/n	n/a
Bluetooth	n/a	Low Energy	n/a

Fig. 1.7 Opta PLC variants (www.arduino.cc)

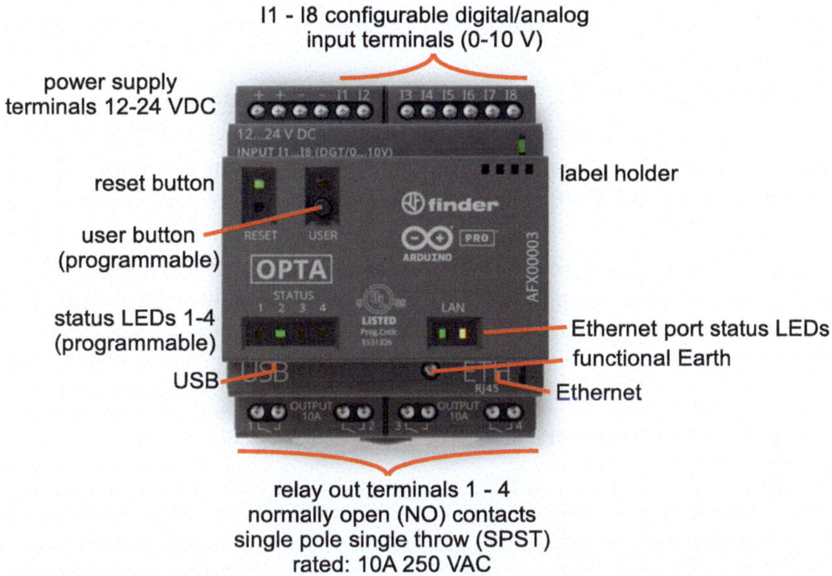

Features
Supply Voltage: 12-24 VDC
Inputs: 8 configurable analog/digital inputs
analog: 0 - 10 VDC, 12-16 bit configurable
digital: 0-24 VDC
Outputs: 4 relay normally open (NO), 10A 250 VAC
fuse: external fuse required
line length (unshielded): 100 m
Processor: Dual-core ST STM32H747XI processor
Dual-core: Arm Cortex-M7 480 MHz core, Arm Cortex-M4 240 Mhz
Program memory: 1 MB RAM
Flash memory: 2 MB
Flash QSPI Interface: 16 MB Quad Serial Peripheral Interface
Internal usage and data logging
USB connectivity
Security: ATECC608B Crypto Microchip

Fig. 1.8 Opta PLC features. Images used courtesy of the Arduino Team (CC BY–NC–SA) (www.arduino.cc)

The three variants (Opta Lite (AFX00003), Opta WiFi (AFX00002), and Opta RS485 (AFX00001) share many common features as shown in Fig. 1.8. Starting in the upper left corner are the DC power supply input terminals for the PLC. The Opta may be powered from a 12–24 VDC power source. We use a 12 VDC supply in our examples. Next is a series of eight configurable input terminals (I1–I8). The terminals may be configured for digital or analog input. There is also an Ethernet 10/100BASE–T port with Local Area Network

(LAN) status indicators. At the bottom of the PLC are four normally open (NO) relay output terminals rated at 250 VAC 10 amps.

The PLC is equipped with a USB–C connector for communication with the host laptop or PC. On the left side of the PLC are programmable, status light emitting diodes (LEDs 1–4) and a programmable user button. Finally, there is a PLC reset button in the upper left corner. The Opta WiFi and RS485 are equipped with additional communication assets we discuss in Chap. 2.

The Arduino Opta variants are programmed using the Arduino IDE or the Arduino PLC IDE. We discuss basic programming techniques using the Arduino IDE in the next section and advanced programming techniques with the Arduino PLC IDE in Chap. 3.

1.11 Getting Started with the Arduino IDE

Most microcontrollers are programmed with some variant of the C programming language. The C programming language provides a nice balance between the programmer's control of the microcontroller hardware and time efficiency in program writing. As an alternative, the Arduino Integrated Development Environment (IDE) provides a user–friendly interface to quickly develop a program or sketch, transform the sketch to machine code, and then load the machine code into the Arduino processor in several simple steps.[4]

The first version of the Arduino IDE was released in August 2005. It was developed at the Interaction Design Institute in Ivrea, Italy to allow the ability to quickly put processing power to use in a wide variety of projects. Since that time, updated versions incorporating new features, have been released on a regular basis (www.arduino.cc).

At its most fundamental level, the IDE is a user–friendly interface to allow one to quickly write, load, and execute code on an Arduino microcontroller or Opta PLC. A barebones program need only consist of a setup() and loop() function. The Arduino IDE adds the other required pieces such as header files and the main program construct. The IDE is written in Java and has its origins in the Processor programming language and the Wiring Project (www.arduino.cc).

In this section we configure the Arduino Opta Lite for basic operation. The Opta Lite is powered from a laboratory power supply set for 12–24 VDC. The Opta Lite is connected to the support PC or laptop via a USB–C cable as shown in Fig. 1.9. The Arduino IDE may be downloaded from the Arduino website's front page at www.arduino.cc. Versions are available for Windows, Mac OS X, and Linux. When the IDE is successfully installed, install the Opta mbed Library using the Library Manager within the Arduino IDE.

Using the Arduino IDE, compile and upload the following sketch from "Getting Started with Opta" tutorial (www.arduino.cc):

---

[4] This section was adapted with permission from: "Arduino I: Getting Started," S. Barrett, Springer Nature, 2020.

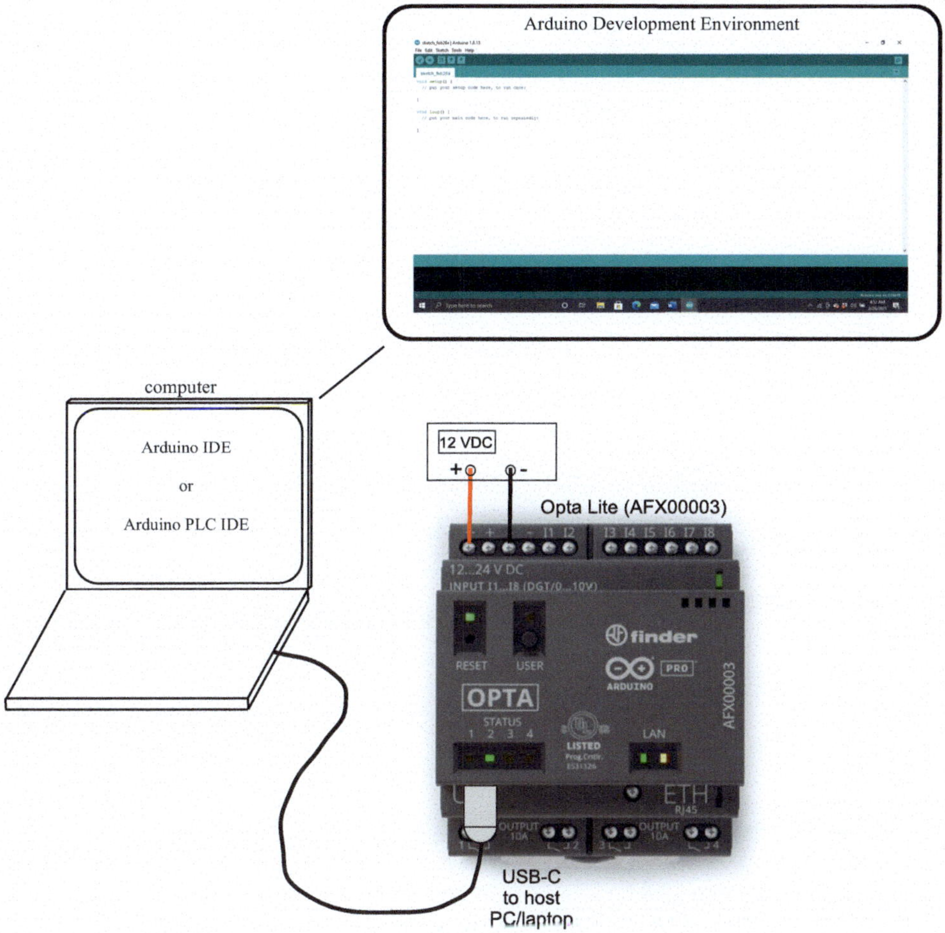

Fig. 1.9 Opta barebones quickstart. Images used courtesy of the Arduino team (CC BY–NC–SA) (www.arduino.cc)

- Under the Tools tab select the evaluation **Board** you are using and the **Port** that it is connected to.
- Upload and execute the program by asserting the "Upload" (right arrow) button.
- When uploaded the four LEDs on the front of the Opta PLC will blink sequentially. Modify the sketch so the LEDs sequentially cycle left to right and then back right to left within a single loop pass.

```
//***************************************************
//Getting Started with Opta
//Name: LED_Blink_Opta
//Purpose: Blink STATUS LEDs on Opta
//author: Arduino
//This is Open Source software.
//****************************************************

void setup()
{
pinMode(LED_D0, OUTPUT);
pinMode(LED_D1, OUTPUT);
pinMode(LED_D2, OUTPUT);
pinMode(LED_D3, OUTPUT);
}

void loop()
{
digitalWrite(LED_D0, HIGH);
delay(100);
digitalWrite(LED_D0, LOW);
delay(100);

digitalWrite(LED_D1, HIGH);
delay(100);
digitalWrite(LED_D1, LOW);
delay(100);

digitalWrite(LED_D2, HIGH);
delay(100);
digitalWrite(LED_D2, LOW);
delay(100);

digitalWrite(LED_D3, HIGH);
delay(100);
digitalWrite(LED_D3, LOW);
delay(500);
}

//***************************************************
```

With the Arduino IDE downloaded and exercised, let's take a closer look at its features.

1.11.1 Arduino IDE Overview

The Arduino IDE is illustrated in Fig. 1.10. The IDE contains a text editor, a message area for displaying status, a text console, a tool bar of common functions, and an extensive menuing system. The IDE also provides a user–friendly interface to the Arduino processor board or PLC which allows for a quick upload of code. This is possible because the Arduino processing boards are equipped with a bootloader program.

A close up of the Arduino toolbar is provided in Fig. 1.11. The toolbar provides single button access to the more commonly used menu features. Most of the features are self–explanatory. As described in the previous section, the "Upload" button compiles your code and uploads it to the Arduino processing board. The "Serial Monitor" button opens the serial monitor feature. The serial monitor feature allows text data to be sent to and received from the Arduino processing board.

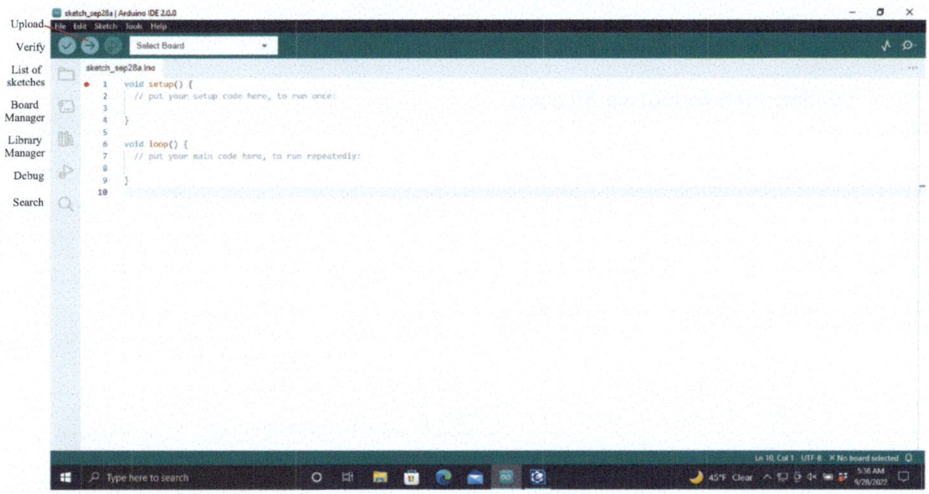

Fig. 1.10 Arduino development environment (www.arduino.cc)

Fig. 1.11 Arduino development environment buttons

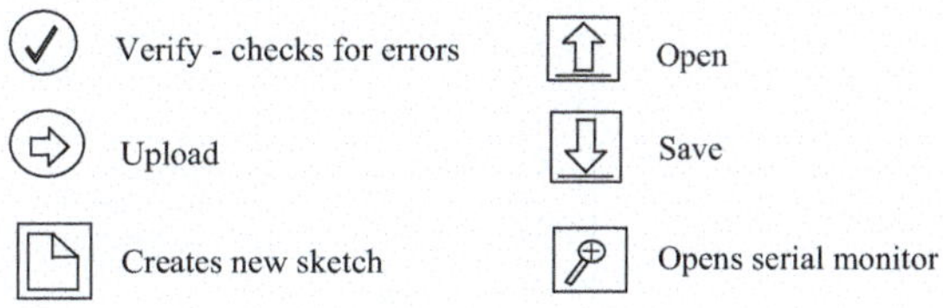

1.11.2 Sketchbook Concept

In keeping with a hardware and software platform for students of the arts, the Arduino environment employs the concept of a sketchbook. An artist maintains their works in progress in a sketchbook. Similarly, we maintain our programs within a sketchbook in the Arduino environment. Furthermore, we refer to individual programs as sketches. An individual sketch within the sketchbook may be accessed via the Sketchbook entry under the file tab.

1.11.3 Arduino Software, Libraries, and Language References

The Arduino IDE has a number of built–in features. Some of the features may be directly accessed via the Arduino IDE drop down toolbar illustrated in Fig. 1.10. Provided in Fig. 1.12 is a handy reference to show the available features. The toolbar provides a wide variety of features to compose, compile, load and execute a sketch.

1.11.4 Writing an Arduino Sketch

The basic format of the Arduino sketch consists of a "setup" and a "loop" function. The setup function is executed once at the beginning of the program. It is used to configure pins, declare variables and constants, etc. The loop function will execute sequentially step–by–step. When the end of the loop function is reached it will automatically return to the first step of the loop function and execute again. This goes on continuously until the program

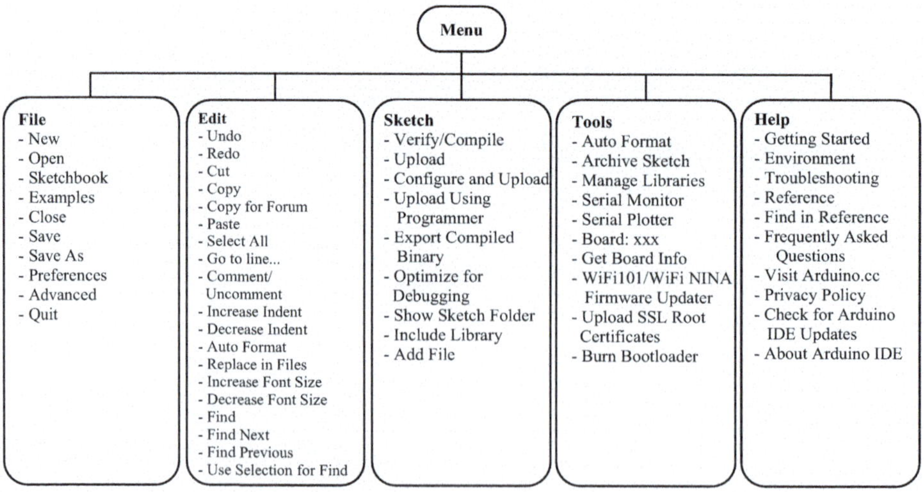

Fig. 1.12 Arduino IDE menu (www.arduino.cc)

is stopped. This serves as a good template for developing an embedded control system. We typically initialize the system (setup) and the continuously monitor and respond to status (loop).

```
//***********************************************************

void setup()
  {
  //place setup code here
  }

void loop()
  {
  //main code steps are provided here
  :
  :

  }

//***********************************************************
```

Even the most complicated sketches follow the basic format of the setup function followed by the loop function. To aid in the development of more complicated sketches, the Arduino IDE has many built–in features that may be divided into the areas of structure, variables and functions. The structure and variable features follow rules similar to the C programming language which is discussed in the text "Arduino II: Systems."[5] The built–in functions consists of a set of pre–defined activities useful to the programmer. These built–in functions are summarized in Fig. 1.13.

There are many program examples available to allow the user to quickly construct a sketch. These programs are summarized in Fig. 1.14. Complete documentation for these programs is available at the Arduino homepage (www.arduino.cc). This documentation is easily accessible via the Help tab on the Arduino Development Environment toolbar. This documentation will not be repeated here. Instead, we refer to these features at appropriate places throughout the remainder of the book. With the Arduino open source concept, users throughout the world are constantly adding new built–in features. As new features are added, they will be released in future Arduino IDE versions. As an Arduino user, you too may add to this collection of useful tools.

---

[5] S.F. Barrett, "Arduino II: Systems," Morgan and Claypool Publishers, 2020.

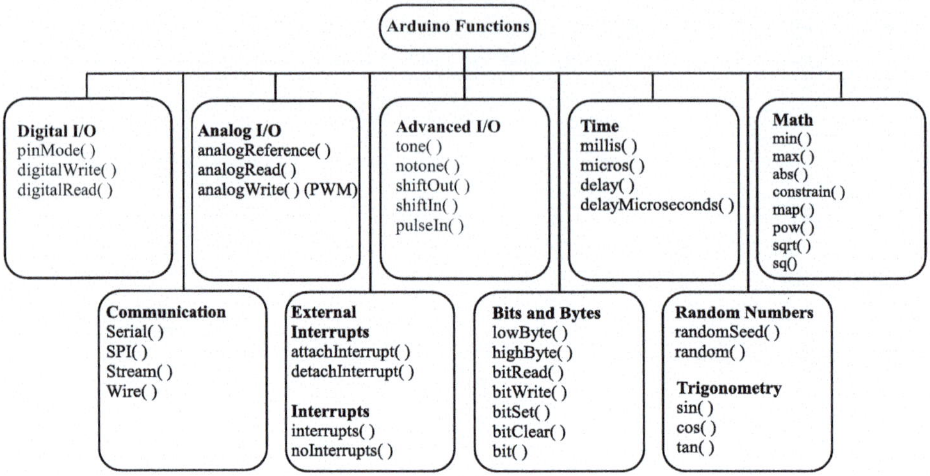

Fig. 1.13 Arduino IDE functions (www.arduino.cc)

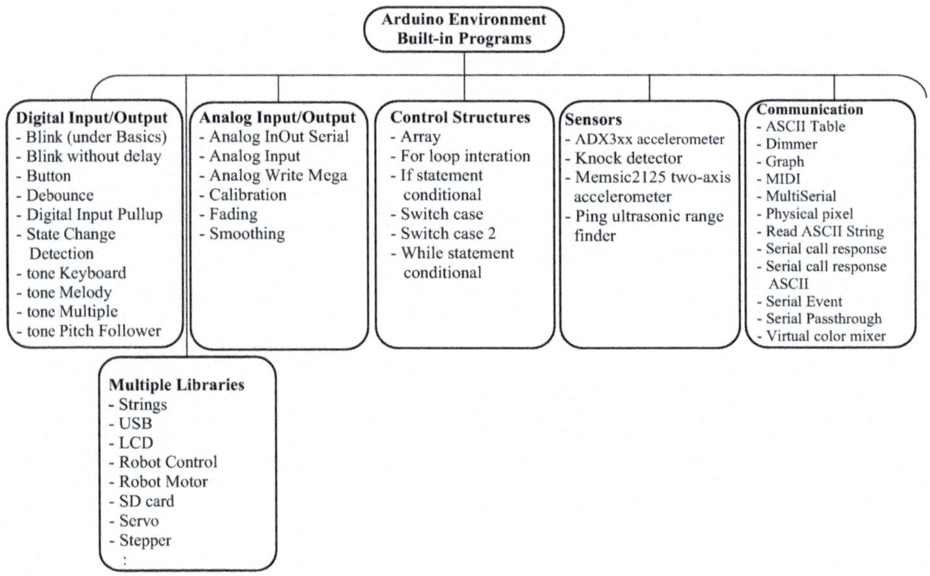

Fig. 1.14 Arduino development environment built–in features (www.arduino.cc)

1.12 Application: Portable Lab Configuration

Provided in Fig. 1.15 is a layout diagram for an Opta PLC panel. Raceway ducts are used to route wiring between components. The DIN compatible components are mounted on standard industrial DIN rails. A 12 VDC, 5A DIN rail power supply (Mean Well MDR-

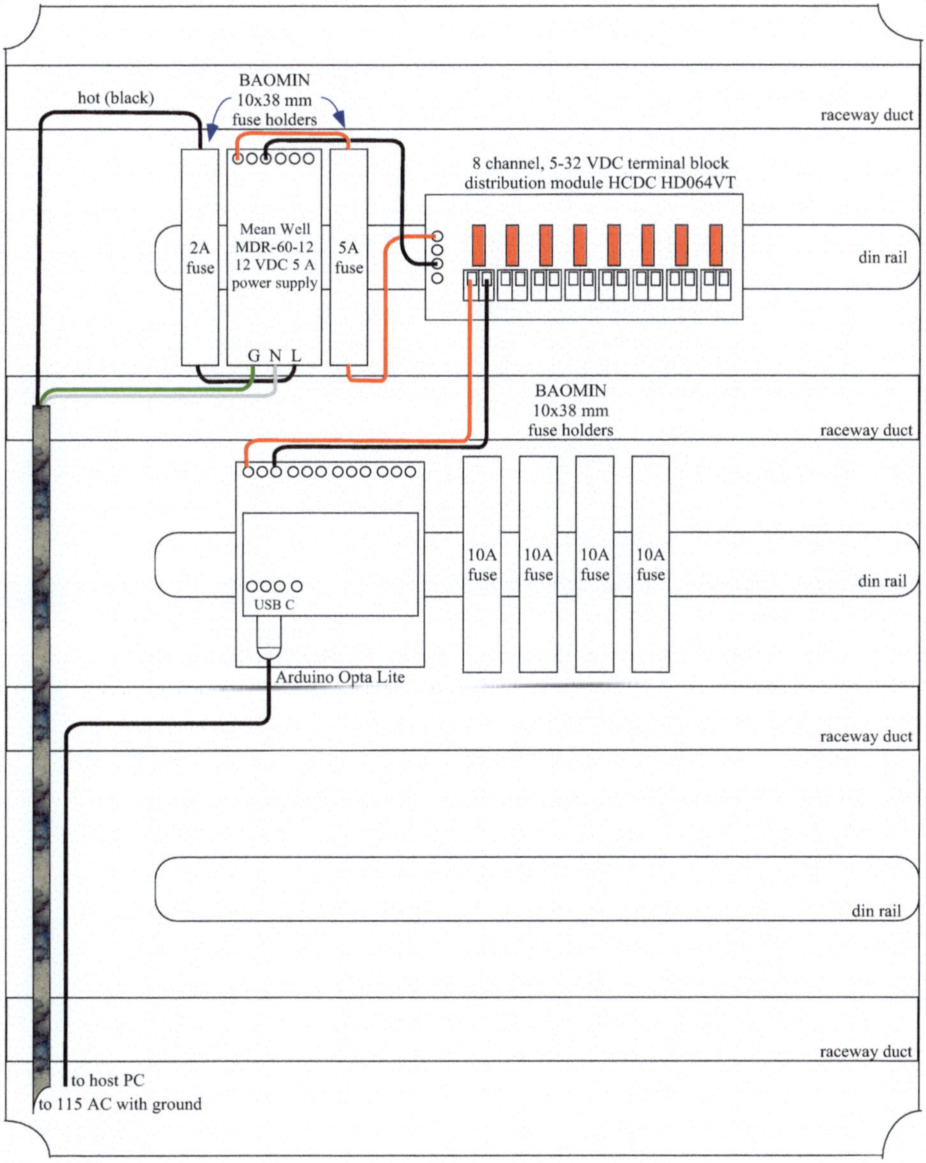

Fig. 1.15 Opta PLC panel

–60\12) powers the Opta PLC. A standard three conductor 115 VAC provides AC power to the 12 VDC supply. The AC input line is fused with a 2A fuse and the 12 VDC output is fused with a 5A fuse. The 10 × 38 mm fuses are housed within a DIN compatible BAOMIN fuse holder.

The 12 VDC supply is distributed to eight channels via the HCDC HD064VT distribution block. Stranded 20 AWG wire is used to connect the components. Block channel 1 is routed to the Arduino Opta Lite power input. The assembled panel is shown in Fig. 1.16.

1.13 Summary

In this chapter we began our exploration of the Operational Technology (OT) world. We started with a basic introduction to the Internet of Things (IoT). Within IoT there is a close relationship between Information Technology (IT) and Operational Technology (OT). We explored this relationship in some detail. The pervasiveness of IoT was then examined in industry or the Industrial Internet of Things (IIoT). We then shifted our focus to OT and basic PLC concepts. We concluded the chapter with an introduction to the Arduino Opta micro PLC.

1.14 Problems

1. Describe different sources of cybersecurity threats.
2. Describe measures to counter cybersecurity threats.
3. Provide a working definition of IoT and IIoT.
4. What is the difference between IT and OT? How are the concepts related.
5. What is a PLC?
6. Describe the PLC scanning process.
7. Provide an IoT model. Describe the interaction between things and applications.
8. What is an "air gap?" Why is it essential for IIoT security?
9. Research and write a short paper on an IIoT security breach. How was the system penetrated? How could the situation been prevented?

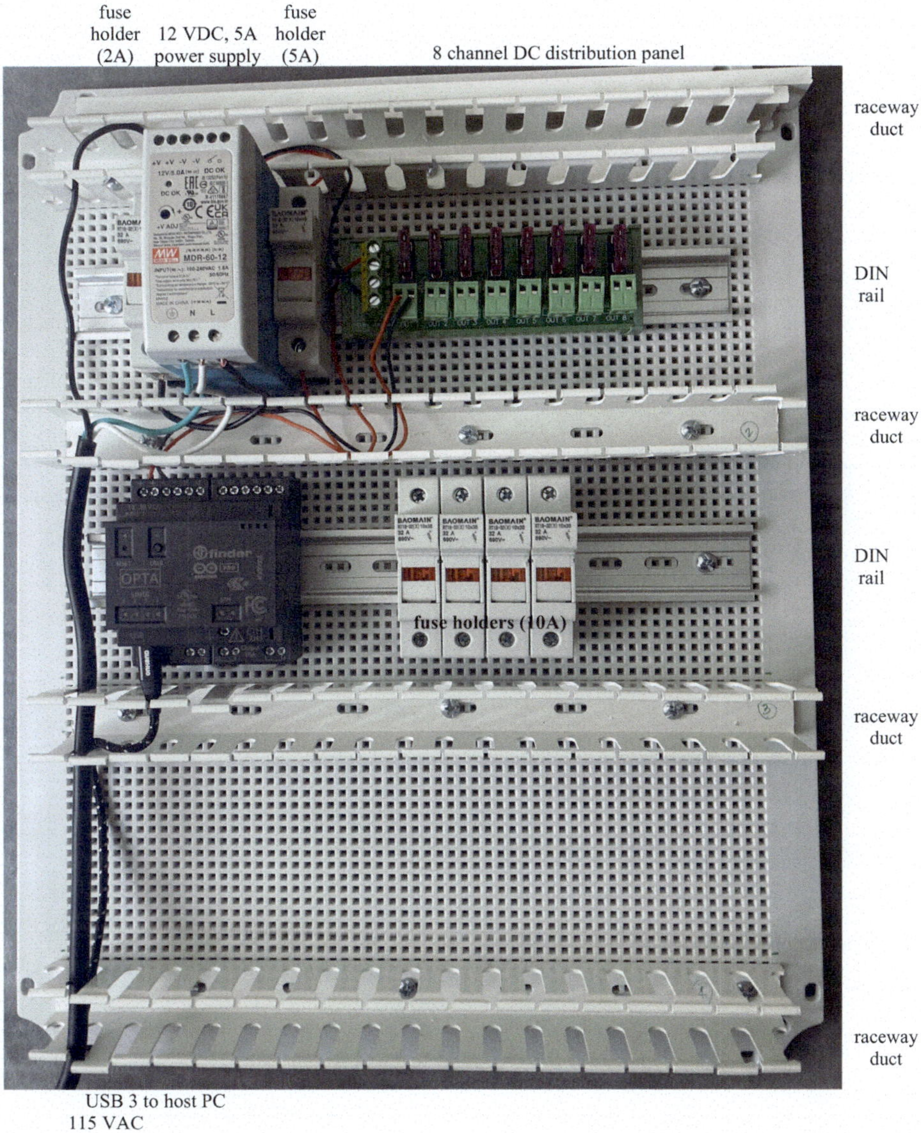

Fig. 1.16 Assembled opta PLC panel

References

Arduino homepage, www.arduino.cc

Greer, C., M. Burns, D. Wollman, E. Griffor (2019) *"Cyber–Physical Systems and Internet of Things*, NIST Special Publications 1900–202, National Institute of Standards and Technology, U.S. Department of Commerce.

Hanes D., G. Salgueiro, P. Grossetete, R. Barton, J. Henry (2017) *IoT Fundamentals–Networking Technologies, Protocols, and Use Cases for the Internet of Things*, Cisco Press.

Kurose, J. and K. Ross (1997) *Computer Networks–A Top–Down Approach*, 7th edition, Pearson Education, Inc.

Levine R. and M. Levine Young (2015) *The Internet for Dummies*, John Wiley and Sons Publishing, Inc.

Lowe, D. (2018) Networking All–In–One for dummies, 7th edition, John Wiley and Sons Publishing, Inc.

Rajkumar, R., I. Lee, L. Sha, J. Stankovic (2010) *Cyber–Physical Systems: The Next Coupling Revolution,* ACM Design Automation Conference, Anaheim, CA,.

Stenerson, J. (2004) *Fundamentals of Programmable Logic Controllers, Sensors, and Communications*, Pearson Prentice Hall.

Z–Wave Alliance, *The Smart Home is Powered by Z–Wave,* z-wavealliance.org.

2 Opta Features

Objectives: After reading this chapter, the reader should be able to do the following:

- Distinguish between different variants in the Arduino Opta micro PLC series;
- Describe common features of the Arduino Opta micro PLC variants;
- Design and implement control circuits employing the common features of the Arduino Opta micro PLC variants;
- Describe and apply features of a specific Arduino Opta micro PLC variant;
- Design a control system employing an Arduino Opta micro PLC; and
- Implement an Arduino Opta micro PLC control system using DIN rail technology.

2.1 Introduction

We begin the chapter with a brief review of the Arduino Opta series of micro PLCs. We explore features common to all three variants and employ them to explore fundamental input/output control concepts. We then examine and apply features specific to a given Arduino Opta variant. Throughout the chapter we provide illustrative examples.

2.2 Arduino Opta Features

As discussed in Chap. 1, the Arduino company has partnered with the Finder company to develop and release a line of basic, yet powerful, Opta programmable logic controllers. The Opta PLCs consist of three variants as shown in Fig. 2.1. The three variants (Opta Lite (AFX00003), Opta WiFi (AFX00002), and Opta RS485 (AFX00001) share many common

S. F. Barrett, *Arduino VII*, Synthesis Lectures on Digital Circuits & Systems,
https://doi.org/10.1007/978-3-031-68609-2_2

Name	Opta Lite	Opta WiFi	Opta RS485
SKU	AFX00003	AFX00002	AFX00001
USB	USB-C	USB-C	USB-C
Ethernet	10/100BASE-T port	10/100BASE-T port	10/100BASE-T port
RS-485	n/a	half-duplex	half-duplex
Wi-Fi	n/a	802.11 b/g/n	n/a
Bluetooth	n/a	Low Energy	n/a

Fig. 2.1 Opta PLC variants (www.arduino.cc)

features. In this section we explore these features. For each feature we provide supporting theory, feature description, and an example using the Arduino IDE. Information provided here is adapted and expanded from *Opta User Manual*, the *Ardunio Opta Collective Data Sheet*, and other sources as referenced.[1]

2.2.1 ST STM32H747XI Dual–Core Processor

The ST STM32H747XI dual–core processor provides the features, subsystems and processing power for the Opta micro PLC line. The processor features an Arm Cortex–M7 core operating at 480 MHz and also an ARM 32–bit Cortex–M4 core operating at up to 240 MHz. The processor is equipped with multiple memory assets including: 1 MB of program memory, 2 MB of Flash memory, and a Flash Quad Serial Peripheral Interface (QSPI) memory. Portions of the QSPI memory is available for manufacturer use and user data logging applications. We explore STM32H747XI features as needed as we investigate Opta PLC features.

2.2.2 Programmable Status LEDs and Push Button

The Opta PLCs are equipped with four programmable status LEDs (1–4) and also a programmable push button. In Chap. 1 we provided a sketch to sequentially illuminate the LEDs. The sketch is provided below.

Within the Arduino IDE the LEDs have been defined as LED_D0 to LED_D3. In the setup portion of the sketch the pins associated with the LEDs are designated output pins.

[1] During the upcoming examples, should the Opta/support computer interface become problematic, the Opta may be reset by depressing the "RESET" pushbutton twice on the Opta front panel.

Within the loop portion of the sketch, the pins are then sequentially turned on (logic HIGH) and off (logic LOW) using the digitalWrite command. Between each command there is a delay of 100 ms to allow viewing of the LED status change.

```
//*****************************************************
//Source: Getting Started with Opta
//Name: LED_Blink_Opta
//Purpose: Blink STATUS LEDs on Opta
//author: Arduino
//This is Open Source software.
//******************************************************

void setup()
{
pinMode(LED_D0, OUTPUT);          //set pins to output
pinMode(LED_D1, OUTPUT);
pinMode(LED_D2, OUTPUT);
pinMode(LED_D3, OUTPUT);
}

void loop()
{
digitalWrite(LED_D0, HIGH);        //set pin HIGH Â- LED ON
delay(100);                        //delay 100 ms
digitalWrite(LED_D0, LOW);         //set pin LOW  Â- LED OFF
delay(100);                        //delay 100 ms

digitalWrite(LED_D1, HIGH);
delay(100);
digitalWrite(LED_D1, LOW);
delay(100);

digitalWrite(LED_D2, HIGH);
delay(100);
digitalWrite(LED_D2, LOW);
delay(100);

digitalWrite(LED_D3, HIGH);
delay(100);
digitalWrite(LED_D3, LOW);
delay(500);
}

//*****************************************************
```

The programmable button is designated as "USER" on the Opta front. Within the Arduino IDE, it is defined as "BTN_USER." The button provides a logic high when not pressed and a logic low when pressed. The following sketch sequentially turns on an additional LED each time the pushbutton is pressed. Once all of the LEDs have turned on the sequence starts over.

The sketch calls the function "changeLights()" to update the LED status based on the value of variable counter. When a function is called, program control is released from the loop program portion to the function. Once the function is complete, program control returns to the main program. Functions may in turn call other functions.

Within the function "changeLights()" a "switch" statement is used to determine appropriate action based on the value of variable "counter." The **switch** statement is used when multiple if–else conditions exist. Each possible condition is specified by a case statement. When a match is found between the switch variable and a specific case entry, the statements associated with the case are executed until a **break** statement is encountered. The alternatives are processed in the order specified by the switch statement. If no match is found, the default case is executed.

```
//*****************************************************
//Source: Getting Started with Opta
//Name: Programmable_Button_Opta
//Purpose: Configures the programmable button to
//control STATUS LED sequence.
//author Arduino
// This is Open Source software.
//*****************************************************

int buttonState = 0;
int counter = 0;

void setup()
{
pinMode(LED_D0, OUTPUT);            //initialize Opta LEDs
pinMode(LED_D1, OUTPUT);
pinMode(LED_D2, OUTPUT);
pinMode(LED_D3, OUTPUT);
pinMode(BTN_USER, INPUT);
}

void loop()
{                                   //check button status
buttonState = digitalRead(BTN_USER);
if(buttonState == LOW)
  {
  if(counter < 4)                   //status of button ctr
    {
    counter++;                      //increment button ctr
    }
  else
```

```
    {
    counter = 0;                          //reset counter
    }
  delay(100);                             //delay 100 ms
  }
changeLights();                           //call function
}

//*****************************************************
//Function to control STATUS LED based on counter value
//*****************************************************

void changeLights()
{
switch(counter)
  {
  case 0: digitalWrite(LED_D0, LOW);
          digitalWrite(LED_D1, LOW);
          digitalWrite(LED_D2, LOW);
          digitalWrite(LED_D3, LOW);
          break;

  case 1: digitalWrite(LED_D0, HIGH);
          break;

  case 2: digitalWrite(LED_D1, HIGH);
          break;

  case 3: digitalWrite(LED_D2, HIGH);
          break;

  case 4: digitalWrite(LED_D3, HIGH);
          break;
  }
delay(100);
}

//*****************************************************
```

2.2.3 Sink and Source Configurations

In the next section we discuss DC inputs and outputs. For DC inputs and outputs we need to be comfortable with the concept of sink and source configurations (Stenerson).

Figure 2.2a illustrates the relationship between two electronic devices with the current flow from one to another. From device 1's point of view, it is sourcing current to device 2. Whereas, from device 2's point of view, device 2 serves as a sink for current from device 1.

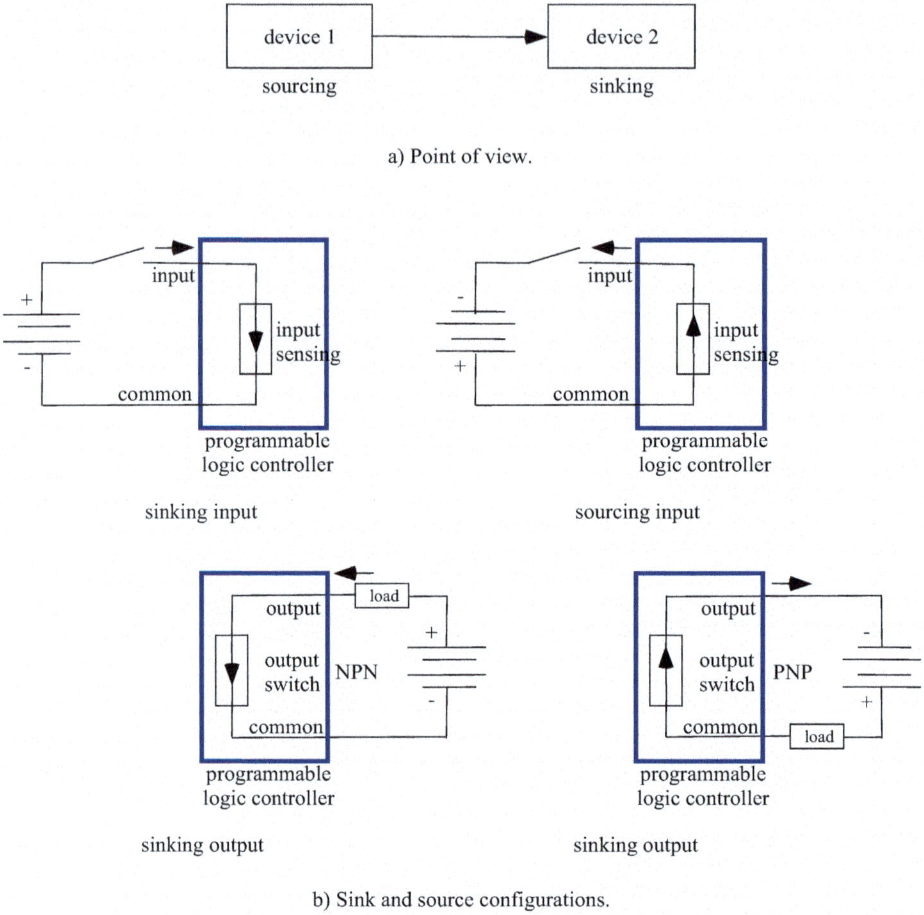

Fig. 2.2 Sink and source configurations (Stenerson)

Field (peripheral) devices may be designed as sink or source devices. The supply voltage for the field devices are typically provided external from the PLC. It is important to correctly configure the field device, power supply, and PLC such that current properly flows in a closed pattern. Figure 2.2b provides common configurations for a PLC in various sink/source configurations. An example of current flow direction with the Opta PLC is provided in Fig. 2.7.

2.2.4 Programmable Analog/Digital Inputs

The Arduino Opta series is equipped with eight inputs designated I1–I8. Each input can be configured as an analog input or a digital input. We discuss each configuration in turn.

2.2.4.1 Analog Inputs

When configured as an analog input, the input signal ranging from 0 to 10 VDC is converted to a corresponding digital value. The corresponding digital value may be configured for 12–16 bits of resolution. A brief introduction to the analog–to–digital (ADC) conversion process follows with an Opta example.

Analog–to–Digital Conversion (ADC) A controller is used to process information from the natural world, use an algorithm to decide on a course of action based on the information collected, and then issue control signals to implement the decision.[2]

Since the information from the natural world, is analog or continuous in nature, and the controller is a digital or discrete based processor, a method to convert an analog signal to a digital form is required. An ADC system performs this task while a digital–to–analog converter (DAC) performs the conversion in the opposite direction.

There are three important processes associated with the ADC process: sampling, quantization, and encoding.

Sampling. Sampling is the process of taking "snap shots" of a signal over time. When we sample a signal, we want to sample it in an optimal fashion such that we capture the essence of the signal while minimizing the use of memory resources. In essence, we want to minimize the number of samples while retaining the capability to faithfully reconstruct the original signal from the samples. Intuitively, the rate of change of a signal determines the number of samples required to faithfully reconstruct the signal, provided that all adjacent samples are captured with the same sample timing intervals.

Harry Nyquist from Bell Laboratory studied the sampling process and derived a criterion that determines the minimum sampling rate for a continuous analog signal. His, now famous, minimum sampling rate is known as the Nyquist sampling rate, which states that one must sample a signal at least twice as fast as the highest frequency content of the signal of interest.

For example, if we are dealing with the human voice signal that contains frequency components that span from about 20 Hz to 4 kHz, the Nyquist sample theorem requires that we must sample the signal at least at 8 kHz, 8000 "snap shots" every second. Figure 2.3 illustrates various sample rates.

[2] The information on analog-to-digital conversion first appeared in "Microcontroller Fundamentals for Engineers and Scientists," Morgan and Claypool Publishers, 2006. It has been adapted with permission. Although first developed for embedded systems design, concepts provided here apply to Opta–based PLC system design.

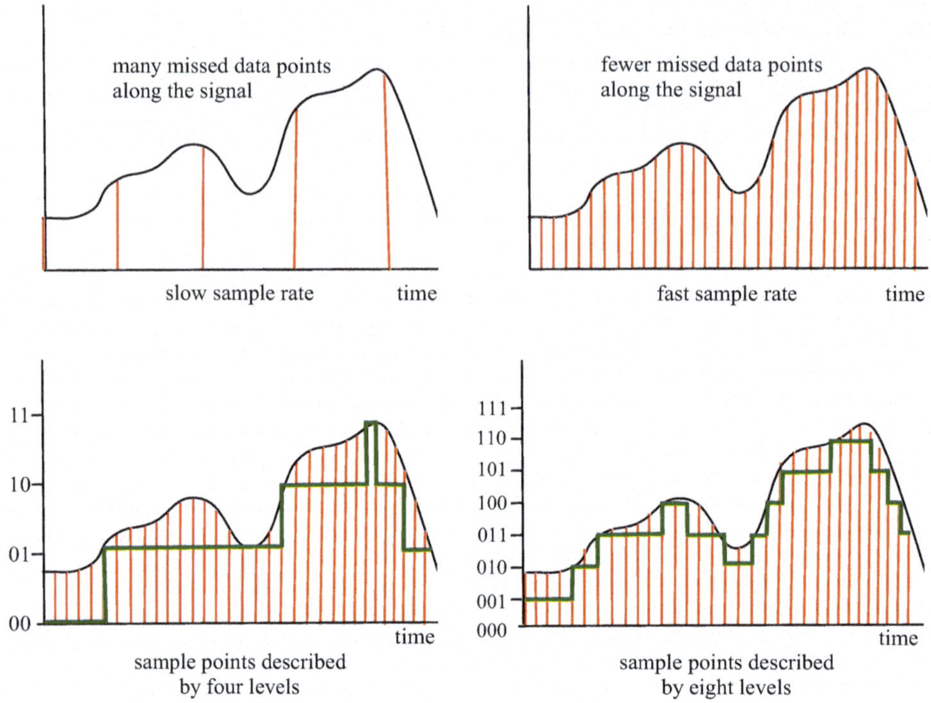

Fig. 2.3 Sampling rate

When a signal is sampled a low pass anti–aliasing filter is employed to ensure the Nyquist sampling rate is not violated. In the example above, a low pass filter with a cutoff frequency of 4 kHz would be used before the sampling circuitry for this purpose.

Quantization. Each digital system has a number of bits it uses as the basic unit to represent data. A bit is the most basic unit where single binary information, one or zero, is represented.

Suppose you have a single bit to represent an incoming signal. You only have two different values, 0 and 1. You may say that you can distinguish only low from high. Suppose you have two bits. You can represent four different levels, 00, 01, 10, and 11. What if you have three bits? You now can represent eight different levels: 000, 001, 010, 011, 100, 101, 110, and 111 as shown in Fig. 2.3. Similar discussion can lead us to conclude that given n bits, we have 2^n unique numbers or levels one can represent.

Figure 2.4 shows how n bits are used to quantize a range of values. In many digital systems, the incoming signals are voltage signals. The voltage signals are first obtained from physical signals (pressure, temperature, etc.) with the help of transducers, such as microphones, angle sensors, and infrared sensors.

The voltage signals are then conditioned to map their range with the input range of a digital system, typically 0–5 VDC for microcontrollers and 0–10 VDC for the Arduino

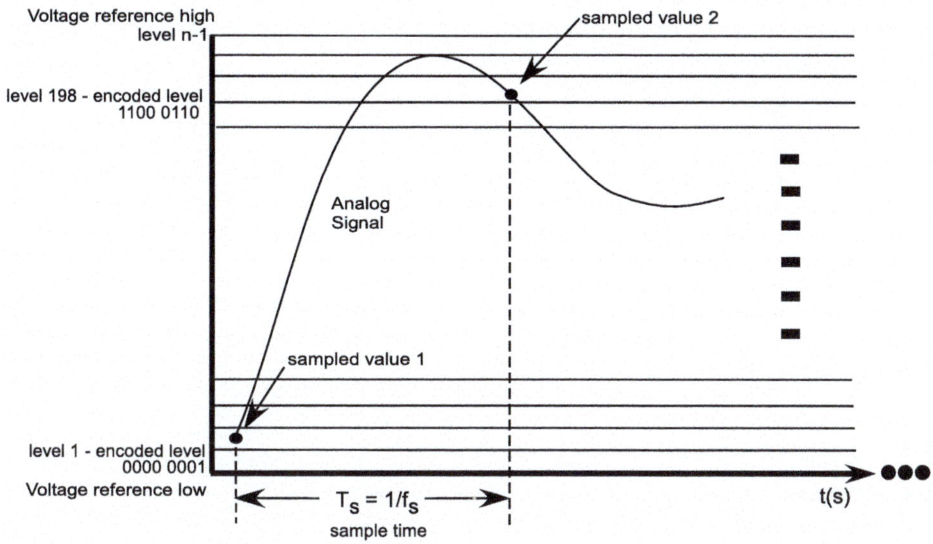

Fig. 2.4 Sampling, quantization, and encoding

Opta PLC. In Fig. 2.4, n bits allow you to divide the input signal range of a digital system into 2^n different quantization levels. As can be seen from the figure, the more quantization levels means the better mapping of an incoming signal to its true value. As the number of bits used for the quantization levels increases for a given input range the "distance" between two adjacent levels decreases accordingly.

Encoding. Finally, the encoding process involves converting a quantized signal into a digital binary value. Suppose again we are using eight bits to quantize a sampled analog signal. The quantization levels are determined by the eight bits and each sampled signal is quantized as one of 256 quantization levels. Consider the two sampled signals shown in Fig. 2.4. The first sample is mapped to quantization level two and the second one is mapped to quantization level 198. Note the amount of quantization error introduced for both samples. The quantization error is inversely proportional to the number of bits used to quantize the signal.

Once a sampled signal is quantized, the encoding process involves representing the quantization level with the available bits. Thus, for the first sample, the encoded sampled value is 0000_0010 (two), while the encoded sampled value for the second sample is 1100_0110 (198). As a result of the encoding process, sampled analog signals are now represented as a set of binary numbers. Thus, the encoding is the last necessary step to represent a sampled analog signal into its corresponding digital form, shown in Fig. 2.4.

Resolution. Resolution is a metric used to quantize an analog signal. Resolution is nothing more than the voltage "distance" between two adjacent quantization levels discussed earlier.

The number of bits used for the quantization is directly proportional to the resolution of a system. In general, resolution may be defined as:

$$resolution = (voltage\ span)/2^b = (V_{ref\ high} - V_{ref\ low})/2^b$$

for the Arduino Opta, the best achievable resolution is:

$$resolution = (10 - 0)/2^{16} = 153\ uV$$

The desired resolution is chosen based on the requirements of the system. The ADC resolution for the Opta PLC can be set from 12 to 16 bits.

Opta ADC example The Arduino Opta input pins are designated I1–I8 on the PLC case. Within the Arduino IDE environment they are designated "PIN_A0" to "PIN_A7." To use the pins as analog inputs, the resolution is set using the "analogReadResolution(insert 12–16 bits)" command.

The maximum allowable voltage to the Arduino Opta PLC is 10 VDC. However, the internal host PLC processor performing the ADC has a maximum allowable voltage of 3 VDC. Therefore, the voltage for ADC conversion provided to the Opta PLC input is scaled internally by a factor of 0.30. When a measurement is taken, the result must be rescaled by this value.

In the following sketch voltage samples provided to Opta PLC inputs I1, I2, and I3 (process inputs A0, A1, A2) are converted, scaled, and displayed to the Arduino IDE serial monitor. The voltages supplied to I1, I2, and I3 are supplied to the Opta from a bank of potentiometers as shown in Fig. 2.5. The Opta readings received need to be calibrated with an external device such as a voltmeter.

```
//*****************************************************
//Source: Getting Started with Opta
//Name: Analog_Inputs_Opta
//Purpose: Test Opta analog pins I1 (A0) to I2 (A1)
//
//author Arduino
//This is Open Source software.
//*****************************************************

void setup()
{
Serial.begin(9600);
analogReadResolution(12);           //set 12 to 16 bits
}

void loop()
{
//Read the input on analog input I1 corresponding to A0:
int sensorValueA0 = analogRead(A0);
```

```
float voltageA0 = sensorValueA0 * (3.0 / 4095.0)/ 0.3;

//Print out value from I1
Serial.print("I1 value: ");
Serial.print(sensorValueA0);
Serial.print(" corresponding to ");

//Print voltage as float with 2 decimal digits
Serial.print(voltageA0, 2);
Serial.println("Volts");

//Read the input on analog input I2 corresponding to A1:
int sensorValueA1 = analogRead(A1);
float voltageA1 = sensorValueA1 * (3.0 / 4095.0)/0.3;

//Print out value from I2
Serial.print("I2 value: ");
Serial.print(sensorValueA1);
Serial.print(" corresponding to ");

//Print voltage as float with 2 decimal digits
Serial.print(voltageA1, 2);
Serial.println("Volts");

//Read the input on analog input I3 corresponding to A2:
int sensorValueA2 = analogRead(A2);
float voltageA2 = sensorValueA2 * (3.0 / 4095.0)/0.3;

//Print out value from I3
Serial.print("I3 value: ");
Serial.print(sensorValueA2);
Serial.print(" corresponding to ");

//Print voltage as float with 2 decimal digits
Serial.print(voltageA2, 2);
Serial.println("Volts");

delay(1000);                        //1 second delay
}

//*****************************************************
```

2.2.4.2 Digital Input/Outputs

In this section we explore the digital inputs and outputs of the Opta PLC.

Digital Inputs The Arduino Opta input pins are designated I1–I8 on the PLC case. Within the Arduino IDE environment they are designated “PIN_A0” to “PIN_A7.”

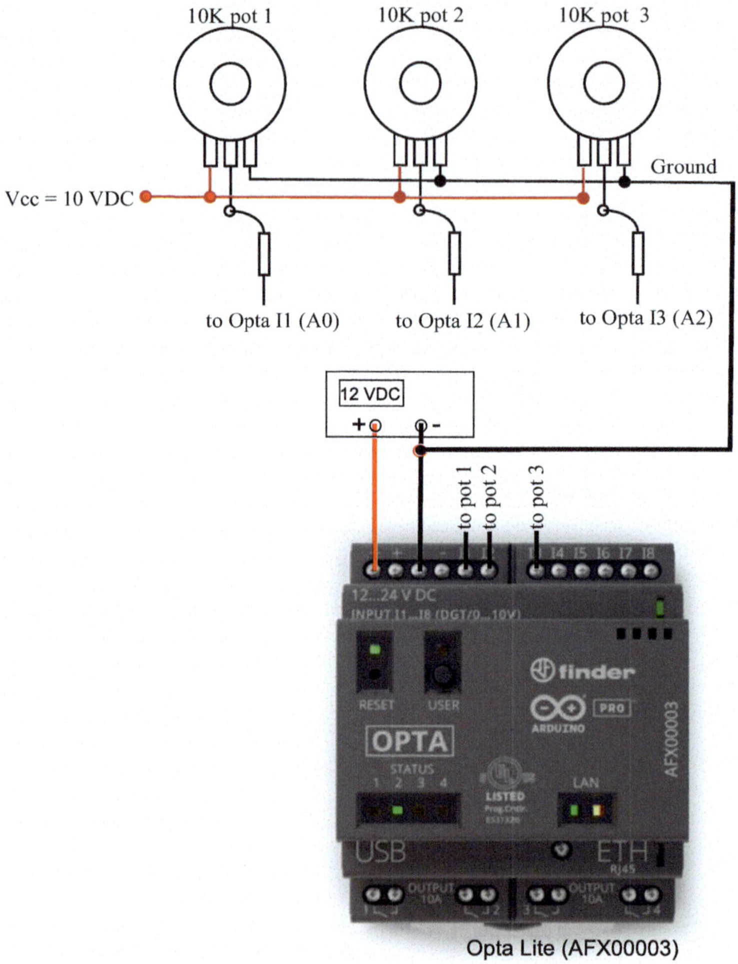

Fig. 2.5 Arduino Opta analog input from potentiometer bank. Images used courtesy of the Arduino team (CC BY–NC–SA) (www.arduino.cc)

The Opta inputs I1–I8 may be configured as digital inputs using:

```
pinMode(pinName, INPUT);
```

A digital input value may range from 0 to 24 VDC. Values less than 4.46 VDC are considered logic low while those greater than 6.6 VDC are considered logic high.

Provided in Fig. 2.6 is a test circuit for digital inputs. Pushbutton tact switches are used between the 12 VDC supply and the Opta input terminals to introduce logic changes to inputs I1, I2, and I3.

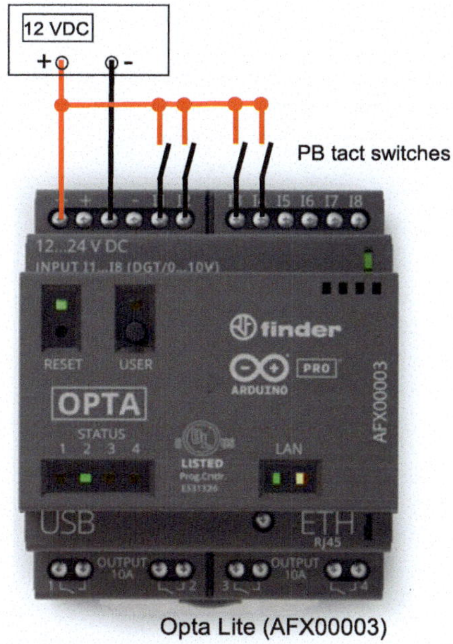

Fig. 2.6 Opta digital inputs. Images used courtesy of the Arduino team (CC BY–NC–SA) (www.arduino.cc)

Mechanical switches do not make a clean transition from one position (on) to another (off). When a switch is moved from one position to another, it makes and breaks contact multiple times. This activity may go on for tens of milliseconds. A processor such as the PLC is relatively fast as compared to the action of the switch. Therefore, the processor is able to recognize each switch bounce as a separate and erroneous transition.

To correct the switch bounce phenomena additional external hardware components may be used or software techniques may be employed. Software switch debouncing is accomplished by inserting a 30–50 ms lockout delay in the function responding to input changes. The delay prevents the processor from responding to the multiple switch transitions related to bouncing.

The following sketch scans for changes in inputs I1–I3. When a switch is depressed, the corresponding Opta status LED is illuminated. Note the use of delays.

```
//***************************************************************
//Opta_input_switch
//
//This is Open Source software.
//***************************************************************

void setup()
{
pinMode(PIN_A0, INPUT);           //Opta inputs I1 to I4
pinMode(PIN_A1, INPUT);
```

```
pinMode(PIN_A2, INPUT);
pinMode(PIN_A3, INPUT);
pinMode(LED_D0, OUTPUT);          //Opta Status LEDs 1 to 4
pinMode(LED_D1, OUTPUT);
pinMode(LED_D2, OUTPUT);
pinMode(LED_D3, OUTPUT);
}

void loop()
{
if(digitalRead(PIN_A0))
  {
  digitalWrite(LED_D0, HIGH);
  delay(100);
  digitalWrite(LED_D0, LOW);
  delay(100);
  }
else if(digitalRead(PIN_A1))
  {
  digitalWrite(LED_D1, HIGH);
  delay(100);
  digitalWrite(LED_D1, LOW);
  delay(100);
  }
else if(digitalRead(PIN_A2))
  {
  digitalWrite(LED_D2, HIGH);
  delay(100);
  digitalWrite(LED_D2, LOW);
  delay(100);
  }
else if(digitalRead(PIN_A3))
  {
  digitalWrite(LED_D3, HIGH);
  delay(100);
  digitalWrite(LED_D3, LOW);
  delay(100);
  }
else
  {
  digitalWrite(LED_D0, LOW);
  digitalWrite(LED_D1, LOW);
  digitalWrite(LED_D2, LOW);
  digitalWrite(LED_D3, LOW);
  delay(100);
  }
}

//************************************************************
```

2.2.4.3 Relay Outputs

The Arduino Opta is equipped with four normally open (NO) mechanical relay contacts. The relay contacts are rated at 250 VAC, 10A. Since they are AC contacts, current can flow in either direction through the relay. The relay outputs are not fused. They may be fused with DIN rail mounted fuses. If you have not read the appendix on safety yet, now would be a good time to do so.

In the following sketch the relays are closed sequentially to illuminate external LEDs and also the Opta STATUS LEDs as shown in Fig. 2.7. Since these are mechanical relays, you can hear as they open and close!

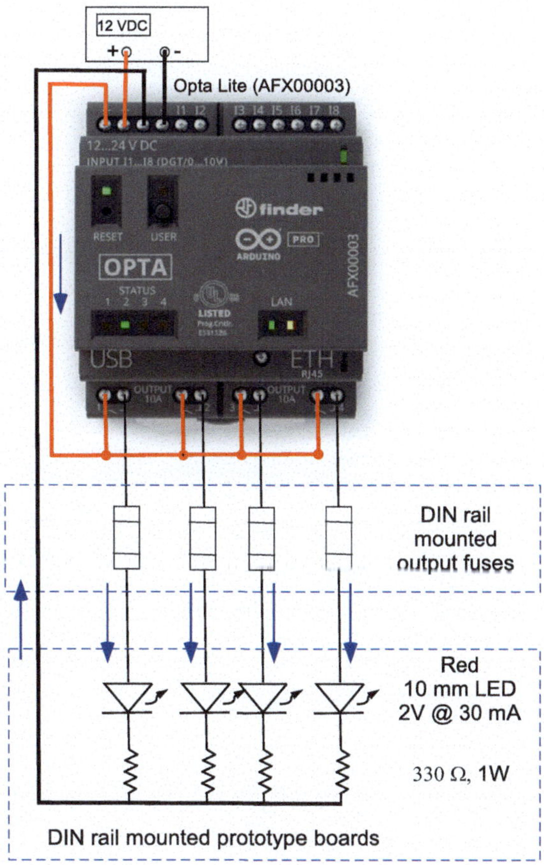

Fig. 2.7 Opta relay outputs. Current flow direction is shown by blue arrows. Images used courtesy of the Arduino team (CC BY–NC–SA) (www.arduino.cc)

```
//****************************************************
//Source: Getting Started with Opta
//Name: Output_Relay_Opta
//Purpose: Test output relays of the Opta
//author Arduino
// This is Open Source software.
//****************************************************

void setup()
{
pinMode(D0, OUTPUT);                //Initialize relay outputs
pinMode(D1, OUTPUT);
pinMode(D2, OUTPUT);
pinMode(D3, OUTPUT);

pinMode(LED_D0, OUTPUT);            //Initialize Opta LEDs
pinMode(LED_D1, OUTPUT);
pinMode(LED_D2, OUTPUT);
pinMode(LED_D3, OUTPUT);
}

void loop()
{
//Closes/opens contact relay 1 and turns on/off LED 1
digitalWrite(D0, HIGH);             //Sets relay 1 on
digitalWrite(LED_D0, HIGH);
delay(1000);
digitalWrite(D0, LOW);              //Sets relay 1 off
digitalWrite(LED_D0, LOW);
delay(1000);

//Closes/opens contact relay 2 and turns on/off LED 2
digitalWrite(D1, HIGH);             //Sets relay 2 on
digitalWrite(LED_D1, HIGH);
delay(1000);
digitalWrite(D1, LOW);              //Sets relay 2 off
digitalWrite(LED_D1, LOW);
delay(1000);

//Closes/opens contact relay 3 and turns on/off LED 3
digitalWrite(D2, HIGH);             //Sets relay 3 on
digitalWrite(LED_D2, HIGH);
delay(1000);
digitalWrite(D2, LOW);              //Sets relay 3 off
digitalWrite(LED_D2, LOW);
delay(1000);

//Closes/opens contact relay 4 and turns on/off LED 4
digitalWrite(D3, HIGH);             //Sets relay 4 on
```

```
digitalWrite(LED_D3, HIGH);
delay(1000);
digitalWrite(D3, LOW);              //Sets relay 4 off
digitalWrite(LED_D3, LOW);
delay(1000);
}

//*****************************************************
```

Digital input/output In the following sketch we combine digital inputs with relay outputs as shown in Fig. 2.8. When a given switch input is depressed, the corresponding relay output closes and illuminates the corresponding external 10 mm LED. Also, the corresponding STATUS LED is illuminated.

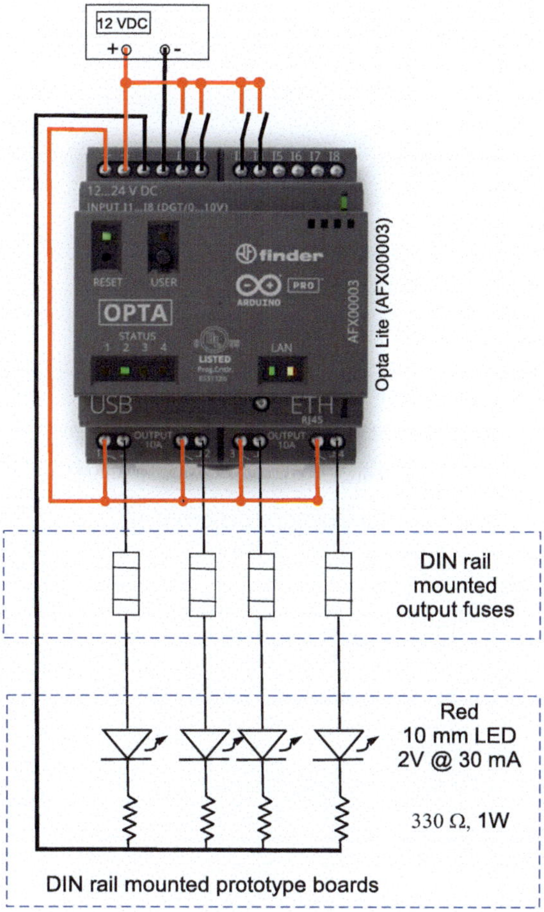

Fig. 2.8 Opta digital inputs with relay outputs. Images used courtesy of the Arduino team (CC BY–NC–SA) (www.arduino.cc)

```
//*****************************************************
//Source: Getting Started with Opta
//Name: Digital_In_Output_Relay_Opta
//Purpose: Test digital inputs with Opta output relays
//author Arduino
//This is Open Source software.
//*****************************************************

void setup()
{
pinMode(PIN_A0, INPUT);             //Opta inputs I1 to I4
pinMode(PIN_A1, INPUT);
pinMode(PIN_A2, INPUT);
pinMode(PIN_A3, INPUT);

pinMode(LED_D0, OUTPUT);            //Opta Status LEDs 1 to 4
pinMode(LED_D1, OUTPUT);
pinMode(LED_D2, OUTPUT);
pinMode(LED_D3, OUTPUT);

pinMode(D0, OUTPUT);                //initialize relays outputs
pinMode(D1, OUTPUT);
pinMode(D2, OUTPUT);
pinMode(D3, OUTPUT);
}

void loop()
{
if(digitalRead(PIN_A0))
  {
  digitalWrite(LED_D0, HIGH);
  digitalWrite(D0, HIGH);           //Sets relay 0 on
  delay(100);
  digitalWrite(LED_D0, LOW);
  digitalWrite(D0, LOW);            //Sets relay 0 off
  delay(100);
  }
else if(digitalRead(PIN_A1))
  {
  digitalWrite(LED_D1, HIGH);
  digitalWrite(D1, HIGH);           //Sets relay 1 on
  delay(100);
  digitalWrite(LED_D1, LOW);
  digitalWrite(D1, LOW);            //Sets relay 1 off
  delay(100);
  }
else if(digitalRead(PIN_A2))
  {
  digitalWrite(LED_D2, HIGH);
```

```
    digitalWrite(D2, HIGH);            //Sets relay 2 on
    delay(100);
    digitalWrite(LED_D2, LOW);
    digitalWrite(D2, LOW);             //Sets relay 2 off
    delay(100);
    }
  else if(digitalRead(PIN_A3))
    {
    digitalWrite(LED_D3, HIGH);
    digitalWrite(D3, HIGH);            //Sets relay 3 on
    delay(100);
    digitalWrite(LED_D3, LOW);
    digitalWrite(D3, LOW);             //Sets relay 3 off
    delay(100);
    }
  else
    {
    digitalWrite(LED_D0, LOW);  digitalWrite(D0, LOW);
    digitalWrite(LED_D1, LOW);  digitalWrite(D1, LOW);
    digitalWrite(LED_D2, LOW);  digitalWrite(D2, LOW);
    digitalWrite(LED_D3, LOW);  digitalWrite(D3, LOW);
    delay(100);
    }
}

//*****************************************************
```

2.2.5 Interrupts

The interrupt system onboard a processor allows it to respond to higher priority events. Appropriate responses to these events are planned, but we do not know when these events will occur. When an interrupt event occurs, the processor will normally complete the instruction it is currently executing and then transition program control to interrupt event specific tasks. These tasks, which resolve the interrupt event, are organized into a function called an interrupt service routine (ISR). Each interrupt will normally have its own interrupt specific ISR. Once the ISR is complete, the processor will return to the main program where it left off before the interrupt event occurred (Fig. 2.9).

The Arduino Development Environment has four built–in functions to support external interrupts (www.arduino.cc).

These are the four functions:

- **interrupts()**. This function enables interrupts.
- **noInterrupts()**. This function disables interrupts.

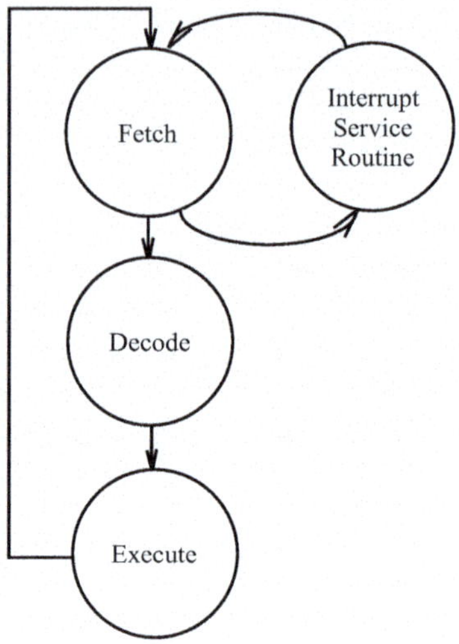

Fig. 2.9 Processor interrupt response

- **attachInterrupt(interrupt, function, mode)**. This function links the interrupt to the appropriate interrupt service routine.
- **detachInterrupt(interrupt)**. This function turns off the specified interrupt.

The Arduino Opta micro PLC series is equipped with interrupts on the USER button (BTN_USER) and all inputs. The **attachInterrupt(interrupt, function, mode)** function is used to link the hardware pin to the appropriate interrupt service pin. The three arguments of the function are configured as follows:

- **interrupt**. Interrupt specifies the interrupt pin.
- **function**. Function specifies the name of the interrupt service routine.
- **mode**. Mode specifies what activity on the interrupt pin will initiate the interrupt: **LOW level on pin, CHANGE in pin level, RISING edge, or FALLING edge.**

Provided below is a template to configure an interrupt.

```
//****************************************************************

void setup()
{
attachInterrupt((BTN_USER), button_ISR, RISING);
}

void loop()
{
```

```
//wait for interrupts

}

//*****************************************************************
//button_ISR: interrupt service routine for (BTN_USER)
//*****************************************************************

void button_ISR(void)
{

//Insert interrupt specific actions here.

}
//******************************************************************
```

2.2.5.1 Foreground and Background Processing

A sequential processor can only execute a single instruction at a time. It processes instructions in a fetch–decode–execute sequence as determined by the program and its response to external events. In many cases, a processor has to process multiple events seemingly simultaneously. How is this possible with a single sequential processor?[3]

Normal processing accomplished by the processor is called foreground processing. An interrupt may be used to periodically break into foreground processing, 'steal' some clock cycles to accomplish another event called background processing, and then return processor control back to the foreground process.

As an example, a processor controlling access for an electronic door must monitor input commands from a user and generate the appropriate pulse width modulation (PWM) signals to open and close the door. Once the door is in motion, the controller must monitor door motor operation for obstructions, malfunctions, and other safety related parameters. This may be accomplished using interrupts. In this scenario, the processor is responding to user input status in the foreground while monitoring safety related status in the background using interrupts as illustrated in Fig. 2.10.

Example: As an example, we configure BTN_USER as an interrupt. During normal operation Opta STATUS LED 1 flashes at one second intervals. When the button is pressed, an interrupt service routine (ISR) is called and sequentially illuminates STATUS LED 2, 3, and 4. Within the ISR multiple NOP instructions are used to generate time delays.

[3] This section is condensed and adapted with permission from "Arduino II: Systems," S. Barrett, Morgan & Claypool Publishers, 2020.

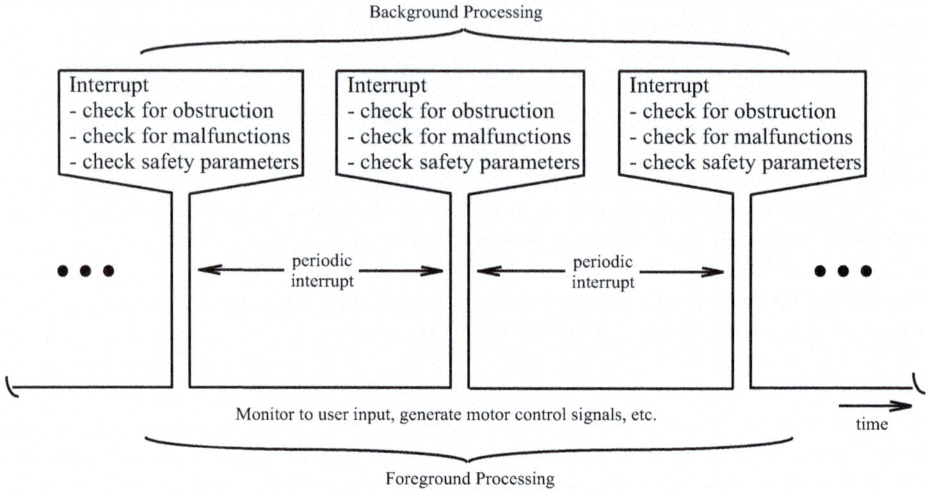

Fig. 2.10 Interrupt used for background processing. The processor responds to user input status in the foreground while monitoring safety related status in the background using interrupts

```
//*****************************************************************
//int_button:  Opta user button (BTN_USER) is configured as
//rising edge interrupt.  In main foreground program, Opta STATUS
//LED1 flashes at 1 second interval.  When button is pressed, ISR
//executes to sequentially illuminate Opta STATUS LEDs 2, 3, 4.
//
////This is Open Source software.
//*****************************************************************

unsigned long int i;

void setup()
{
pinMode(LED_D0, OUTPUT);            //Opta Status LEDs 1 to 4
pinMode(LED_D1, OUTPUT);
pinMode(LED_D2, OUTPUT);
pinMode(LED_D3, OUTPUT);
pinMode(BTN_USER, INPUT);
attachInterrupt(BTN_USER, background, RISING);
}

void loop()
{                                   //foreground processing
digitalWrite(LED_D0, HIGH);         //Opta STATUS LED1 on
delay(500);                         //500 ms delay
digitalWrite(LED_D0, LOW);          //Opta STATUS LED1 off
delay(500);                         //500 ms delay
```

```
}

void background()                    //Interrupt Service Routine
{                                    //background processing
digitalWrite(LED_D0, LOW);           //Opta STATUS LED1 off
digitalWrite(LED_D1, HIGH);          //Opta STATUS LED2 on
for (i=0; i<=240000000; i++)         //1s delay 240e6 clock cycles
  {
  asm("nop");                        //1 clock cycle
  }

digitalWrite(LED_D1, LOW);           //Opta STATUS LED2 off
digitalWrite(LED_D2, HIGH);          //Opta STATUS LED3 on

for (i=0; i<=240000000; i++)         //1s delay 240e6 clock cycles
  {
  asm("nop");                        //1 clock cycle
  }

digitalWrite(LED_D2, LOW);           //Opta STATUS LED3 off
digitalWrite(LED_D3, HIGH);          //Opta STATUS LED4 on

for (i=0; i<=240000000; i++)         //1s delay 240e6 clock cycles
  {
  asm("nop");                        //1 clock cycle
  }

digitalWrite(LED_D3, LOW);           //Opta STATUS LED4 off

}

//*****************************************************************
```

2.2.6 Internet Connectivity

All three of the Arduino Opta variants are equipped with internet features including the capability to establish Ethernet connectivity with other devices. In this section we begin with fundamental internet concepts and also an Ethernet example. In the section following we provide for internet connectivity via WiFi using the Opta WiFi variant.

2.2.6.1 A Big Picture of the Internet

From its early beginnings in the late 1960s to today, the internet has become ubiquitous (found everywhere) in every facet of our lives. A few examples where the internet has

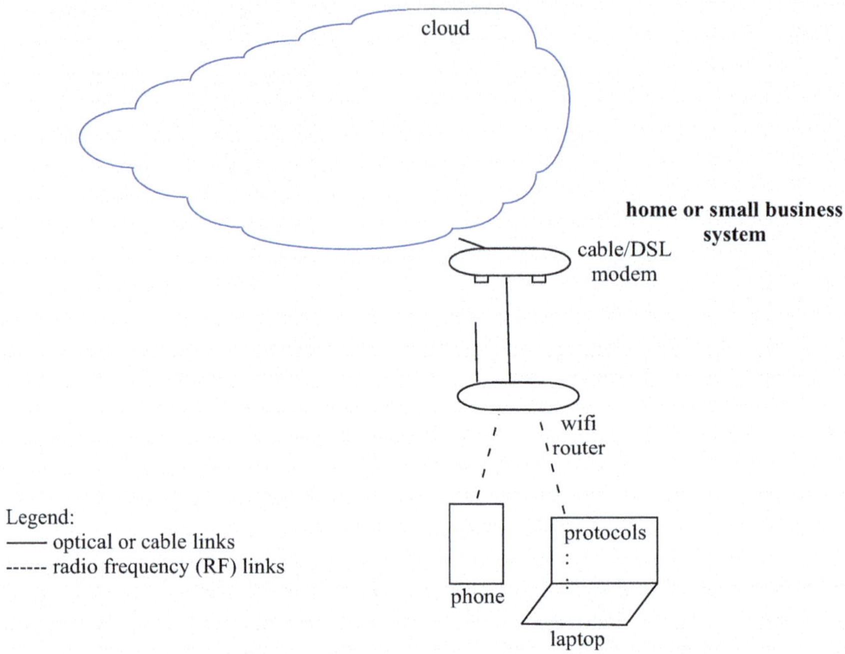

Fig. 2.11 Internet at home or a small business configuration

become prevalent include industry, agriculture, energy production, education, healthcare, entertainment, manufacturing, retail, communications, and many other areas. References for the following internet sections are from (Hanes, Lowe, and Null) and others as cited.[4]

Many (including the author) take a safe, secure, and reliable internet for granted. Portions of the internet, consisting of a global network of interconnected computers, are referred to as the "cloud." In this section we examine connections between computers that comprise the internet in a home and work environment and then examine what is inside the cloud.

Figure 2.11 provides a typical internet connection found in a home or small business such as a cafe or small store. A cable or digital subscriber line (DSL) modulator/demodulator (modem) provides a connection to the internet. The cable/DSL modem is provided by an Internet Service Provider (ISP) when you subscribe to their internet connection service. The connection between the cable/DSL modem and the ISP provider may be a combination of copper cable, optical fiber, and wireless radio frequency connection links.

With internet service available via the cable/DSL modem, a WiFi router is used to establish a wireless local area network (WLAN) within your home or small business. The WLAN serves as an internet access point to a broader area using radio frequency (RF) signals operating at 2.5 or 5 GHz. The link between the cable/DSL modem to the WiFi router is

[4] Portions of this section are adapted with permission from "Arduino III: Internet of Things," S. Barrett, Springer, 2021.

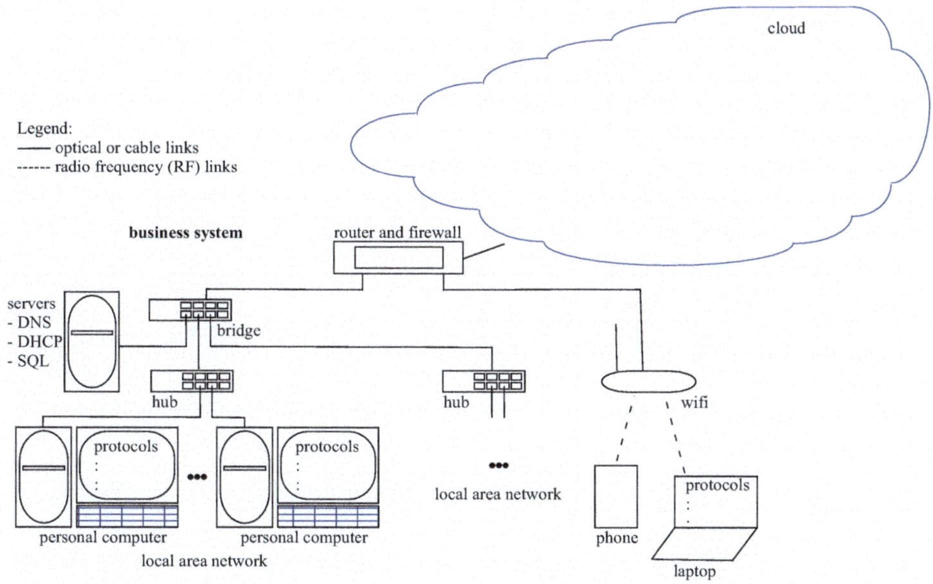

Fig. 2.12 Internet at a large business or university

via a cable. Wireless devices such as a cell phone or a laptop within range of the WiFi router are able to access the internet with proper credentials (i.e. WLAN password and ISP subscription). The WiFi router range may be extended using a WiFi range extender.

Internet service typically found in a larger business or a university is shown in Fig. 2.12. Internet service is provided to the organization via an ISP. Connection is made to the ISP via a firewall. The firewall provides protection from internet hazards outside the organization. It may also be used to limit outgoing information from the organization (e.g. sensitive company information, classified material, etc.). There is also a router at the organization portal. It is used to route internet message traffic to its next destination.

Individual computers are provided access to the internet via a cabled connection to a hub. The hub is used to connect computers into a local area network (LAN) sharing the same location or similar function such as an academic department or company section. Multiple LANs are then connected via a bridge. The bridge will also provide connection to a series of local servers such as a Domain Name System (DNS) server, a Dynamic Host Configuration Protocol (DHCP) server, a Structured Query Language (SQL) database server, and a mail server (among others). WiFi access may also be provided by a WiFi router as previously described.

2.2.6.2 Internet Cloud

Figure 2.13 shows the configuration of the internet cloud. The cloud contains the global connection of multiple internet service providers. Regional internet service providers share internet traffic via a metropolitan area exchange (MAE). The regional network ISPs connect to a network service provider who are in turn connected to other network service providers via network access points. Connectivity across the globe is provided by submarine optical fiber cables spanning the oceans. Internet access may be provided to remote areas via balloon borne network access points. The overall result is a global network of interconnected computers for the open exchange of information.

Aside from the internet hardware components, there are internet protocols and applications used to ensure reliable and compatible communications from one location to another. As an example, when you are checking your favorite news website or sending an e–mail to

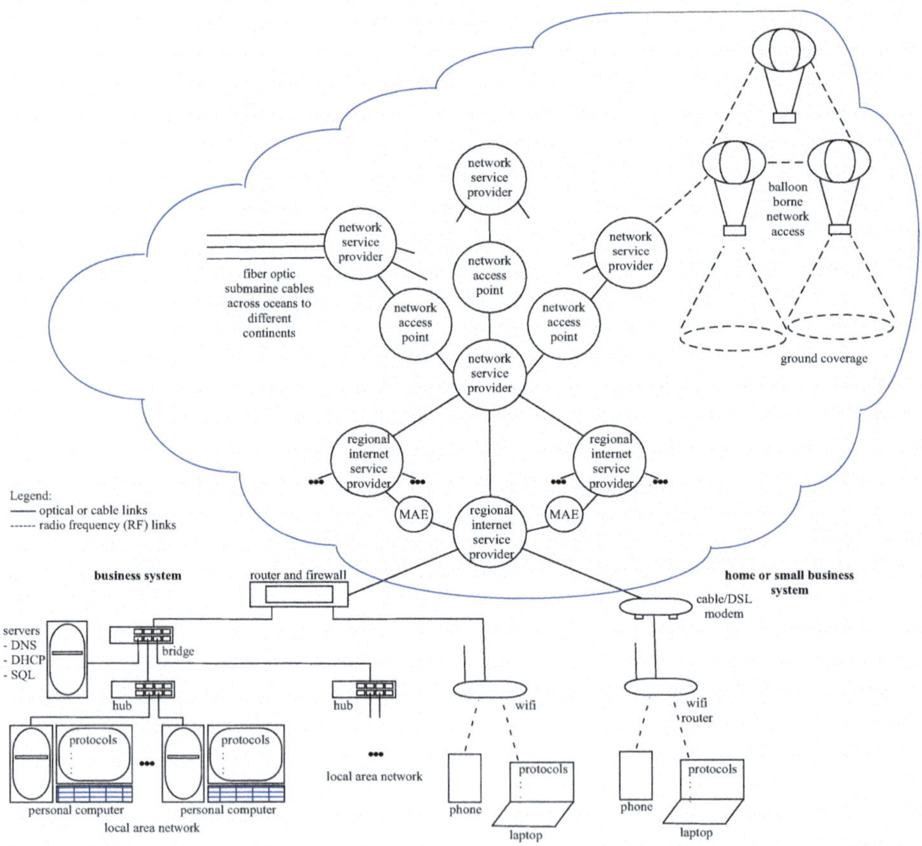

Fig. 2.13 The cloud

a friend, the specific application you are using along with the computers operating system has built in features to interact with the internet to accomplish your desired task.

The application and the operating system apply the different layer activities to transmit information over the internet. For example, the e–mail message you are sending is broken into packets of information, provided source and destination IP addresses, and converted to an electronic signal. The packets are then routed to the destination computer via a series of internet hops directed by routers along the path. At the destination the received packets are reassembled and the protocol steps are applied in reverse order.

2.2.6.3 Internet Protocol Models

There are two different models of protocol stacks commonly used within the internet community. A protocol is a standardized set of rules and procedures. The layered protocol models provide guidelines on how data is processed within applications and prepared for transmission over the internet. Both protocols were developed in the early 1980s.

The International Organization for Standardization (ISO) developed the seven layered Open Systems Internet protocol stack ISO/OSI reference model shown in Fig. 2.14a. The adjacent layers interact with one another in a given system while similar layers interact with one another in different systems.

The transmission control protocol/internet (TCP/IP) maps into several layers of the OSI model as shown in Fig. 2.14a. Figure 2.14b shows the four layers of the TCP/IP model: application, transport, internet, and link layer.

The data to be shared with another computer resides within the application layer. The data is divided into packets. As the data packet is processed through each layer of the sending computer, additional header and footer information is appended to the data payload. The IP address allows the sender and receiver to find one another on the internet (Leiden, Lowe, Null).

2.2.6.4 Internet Addressing Techniques

In this section we discuss the importance of and techniques used to address network assets. We begin with IP addressing and packet headers. There are two different versions of the IP header: IPv4 and IPv6 as shown in Fig. 2.15.

IPv4 header The earlier IPv4 header version consists of 24 bytes followed by the data payload. The overall packet datagram must be at least 40 bytes. The IPv4 header consists of the following fields:

- version: IP protocol version. For IPv4 this field is set to $(0100)_2$.
- header length: specified as 32–bit words.
- type of service: specifies priority from low (000) to critical (101)
- total length: specifies total length of the datagram packet in bytes.

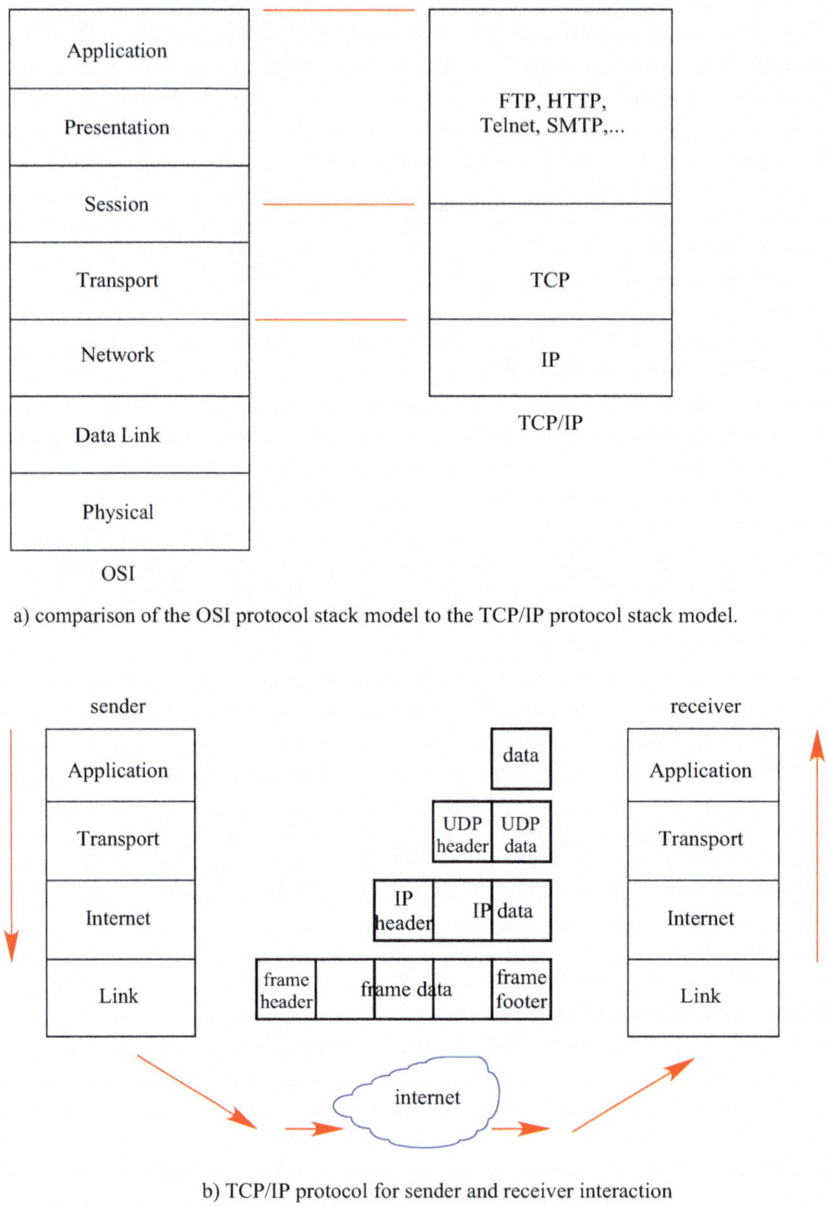

Fig. 2.14 OSI versus TCP/IP (Leiden, Lowe, Null)

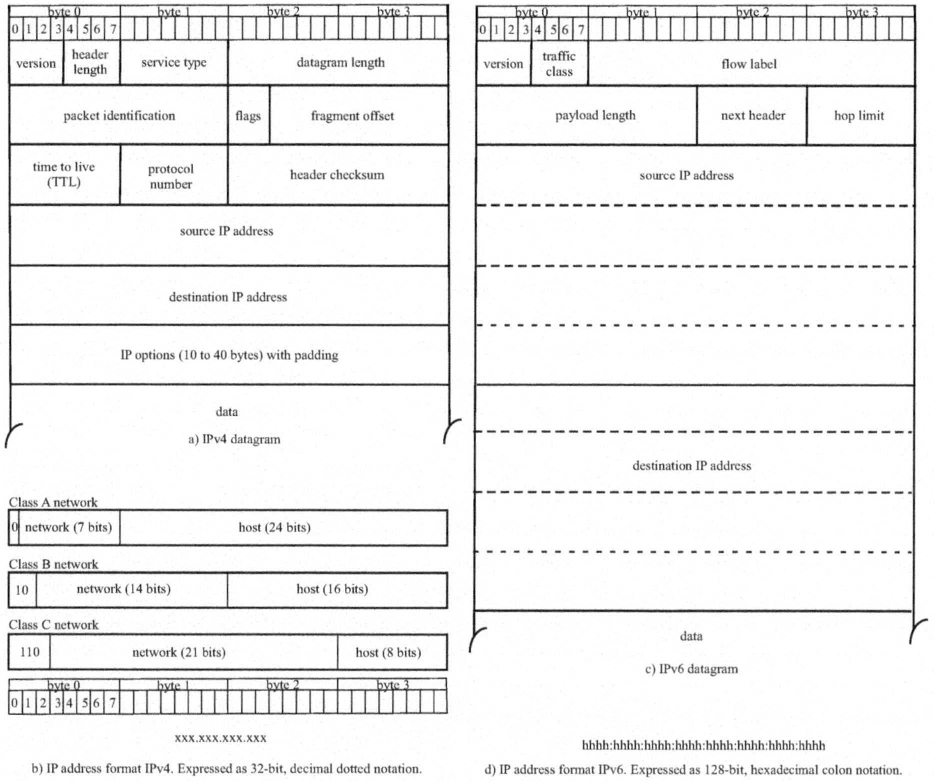

Fig. 2.15 IPv4 and IPv6 (Hanes, Null)

- packet ID: the packet is assigned a unique serial number
- flags: indicates whether a larger packet can be broken into smaller packets.
- fragment offset: provides fragment location within packet.
- time to live: determines number of internet hops are allowed from source to destination.
- protocol number: indicates the type of protocol associated with the data: 0–reserved, 1–internet control message protocol (ICMP), 6–transmission control protocol (TCP), or 17–user datagram protocol (UDP).
- header checksum: holds the calculated checksum of the header.
- provides source and destination address of packet. Note each address is 32 bits in length. A unique address or source ID is provided to each source computer on the internet. The allocation of addresses is coordinated by the Internet Corporation for Assigned Names and Numbers or ICANN (www.ICANN.org).
- IP options: provides additional control information.

The IPv4 protocol uses the 32–bit IP address configuration as shown in Fig. 2.15b. The first several bits of the address indicates the network class: Class A for large networks (0), Class B for medium sized networks (10), and Class C for smaller networks (110). The remaining IP address bits are partitioned to select a network and a specific host. The IP address is expressed as a 32–bit dotted decimal notation value (xxx.xxx.xxx.xxx). Each byte in the address is specified by its decimal equivalent (xxx) and has a value ranging from 0 to 255 (Hanes, Lowe, Null).

CIDR addressing To provide additional addressing flexibility a Classless Inter–Domain Routing (CIDR) addressing scheme was developed. The CIDR scheme provides for flexible subnet addressing within the IPv4 protocol. The format of the CIDR address is provided in Fig. 2.16a. A subnet is a smaller network within one of the network classes: A, B, or C.

Recall the IPv4 protocol provides for a 32–bit IP address. The CIDR protocol allows the address to be partitioned into a number of address bits allocated to identify the network and the remaining bits allocated to identify a specific host (computer) within the network. A slash character (/) follows the address with a decimal number. The decimal number indicates the number of logic ones in the subnet mask.

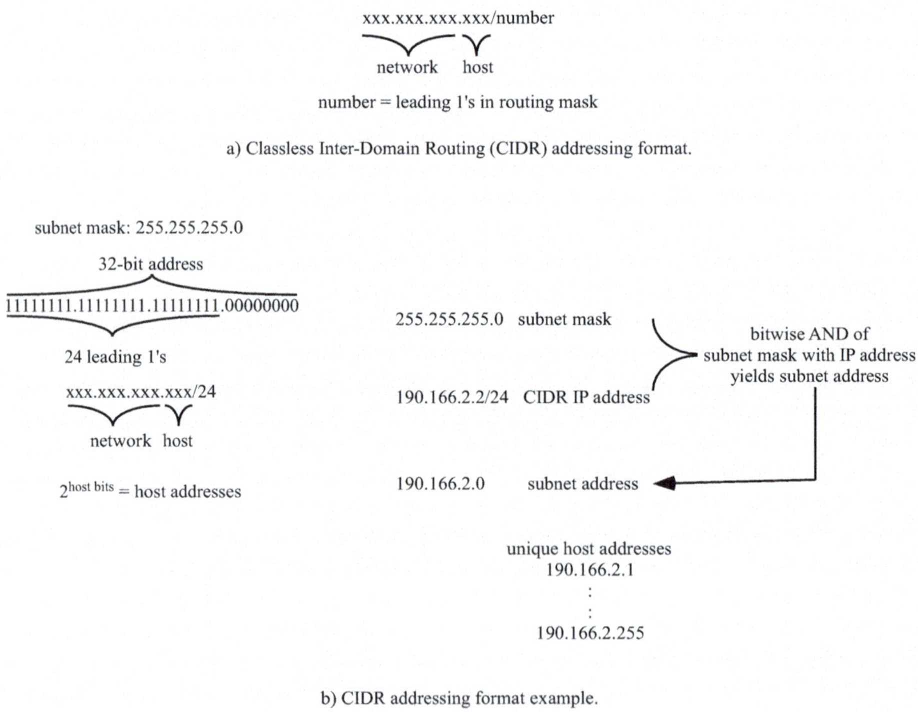

Fig. 2.16 CIDR addressing (www.IETF.org)

In the example provided in Fig. 2.16b left an IP address has been partitioned such that 24 bits will be used to identify the specific network while the remaining bits eight bits will be used to specify a specific computer within the network. This partition allocation results in a subnet mask of 255.255.255.0 or expressed in separated binary as $(11111111.11111111.11111111.00000000)_2$. Since there are 24 bits in the subnet mask containing leading ones, a /24 is appended to the IP address to communicate the partition information. The eight bits allocated for host or computer addressing will provide for $2^{host\ bits}$ or 256 unique addresses.

In Fig. 2.16b right, the IP address (190.166.2.2./24) has been partitioned into 24 bits for the subnet address and the other eight bits for specific computer addressing. This partition results in the subnet mask 255.255.255.0. When this mask is logically ANDed with the IP address the resulting subnet address 190.166.2.0 results. The specific computers within the subnet will be addressed beginning at 190.166.2.1 and ending at 190.166.2.255 (www.IETF.org).

IPv6 header In the mid–1990s the IPv6 was released by the Internet Engineering Task Force (IETF). The IPv6 protocol provides for a longer 128–bit IP address space. The different fields within the IPv6 header specify:

- version: IP protocol version. For IPv6 this field is set to $(0110)_2$.
- traffic class: will specify different priority.
- flow label: will specify the type of communication in progress.
- payload length: expressed in bytes
- next header: specifies if additional header information is provided in the payload.
- hop limit: will allow up to 256 hops from source to destination.
- source and destination addresses: 128–bits each.

The IETF developed a logical, methodical method of assigning IPv6 addresses, the Aggregatable Global Unicast Address Format, as shown in Fig. 2.17. The 128–bit address is partitioned into different fields to specify top, next, and site–level aggregation representing for an example a country, a company within the country, and networks within the company, respectively. The 64–bit interface ID is a combination of the host device MAC address and information from a nearby router. The 128–bit IPv6 address is specified as eight 16—bit values expressed in hexadecimal and separated by colons (hhhh:hhhh:hhhh:hhhh:hhhh:hhhh:hhhh:hhhh) (Hanes, Lowe, Null).

MAC address The host device's Medium Access Control or MAC address is a 48–bit device specific address as shown in Fig. 2.18. The address is partitioned into six different bytes. The address specifies the Organizational Unique Identifier (OUI) and the Network Interface Controller (NIC) identifier. The NIC provides the interface between the host computer or device and the internet. The MAC addressing scheme allows each NIC to have a unique address (Hanes, Lowe, Null).

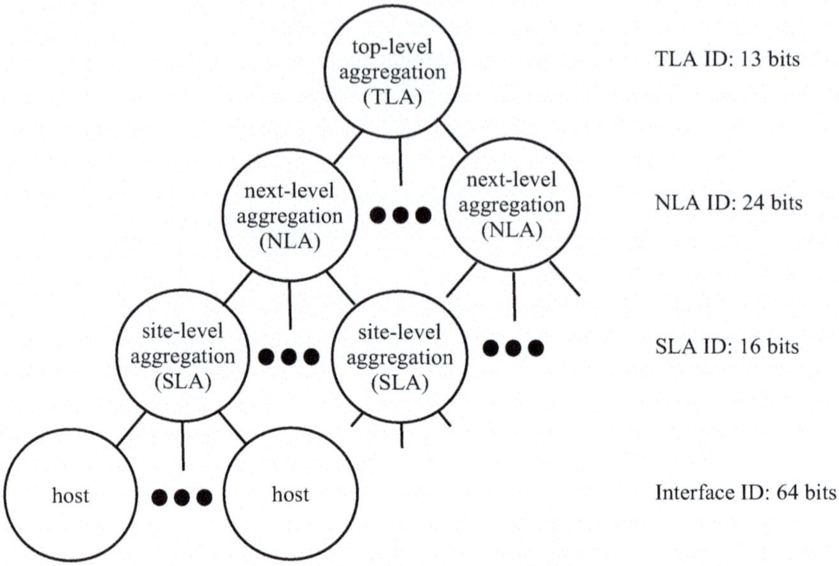

a) Aggregatable Global Unicast Address

Prefix 001 3 bits	TLA ID 13 bits	Reserved 8 bits	NLA ID 24 bits	SLA ID 16 bits	Interface ID 64 bits

b) Aggregatable Global Unicast Address Format

Fig. 2.17 IPv6 addressing (Hanes, Lowe, Null)

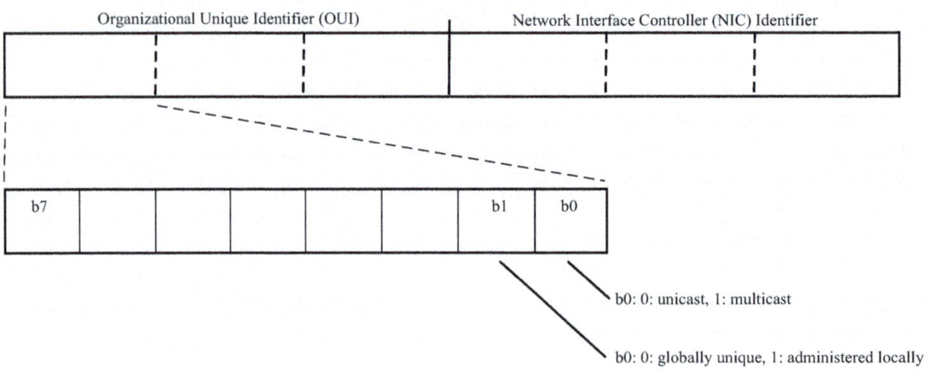

Fig. 2.18 MAC address (Hanes, Lowe, Null)

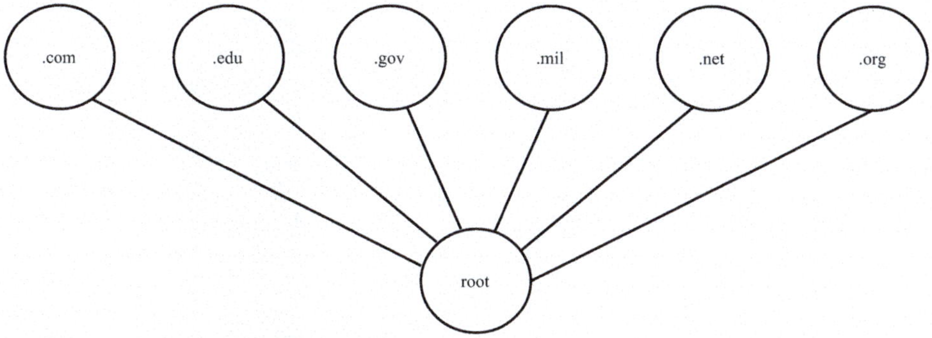

Fig. 2.19 Common URL domains (Shuler)

DNS and URL addressing Rather than memorize the IP address for individual networks and computers, descriptive, user–friendly names may be assigned using Domain Name System (DNS) techniques. The DNS serves as a distributed directory of named network assets. The directory is stored on a number of DNS servers throughout the internet. You may apply for a DNS name from a DNS provider. The provider will determine if the name is available for use. The Internet Corporation for Assigned Names and Numbers (ICANN) coordinates the use of DNS names across the globe (Shuler).

To completely specify the location of a computer on the internet, a Uniform Resource Locator (URL) address is used. The URL address consists of three parts: the protocol identifier, the DNS name, and the domain name.

A common protocol is "http." The protocol type is followed by ": $//www.$." The next portion of the URL address is the DNS name followed by the domain (e.g. .edu, .org, etc.). URL domains are shown in Fig. 2.19. As an example, the website address for the main Arduino site is: http://www.arduino.cc. The ".cc" is a variant of the ".com." domain.

2.2.7 Ethernet 10/100BASE–T Port

Each Arduino Opta variant provides access to Ethernet features via a standard RJ45 connector. The ethernet is a method of connecting computers within a Local Area Network (LAN). The 10BASE–T designator indicates a maximum transmission rate of 10 Mbps per second (10) of baseband signal transmission (BASE) with twisted pair cable (T). The 100BASE–T's maximum transmission rate is 100 Mbps per second.

Example: Ethernet 1. For this demonstration, an Ardunio Opta is connected to a home router via an RJ45 cable as shown in Fig. 2.20. The Opta is programmed using a laptop/PC. In the sketch the Opta is configured as a client. It will connect to a server configured computer via the internet and request the service provided. In this specific example the server is located at www.ip-api.com. This website hosts an IP Geolocation Application Programming Interface

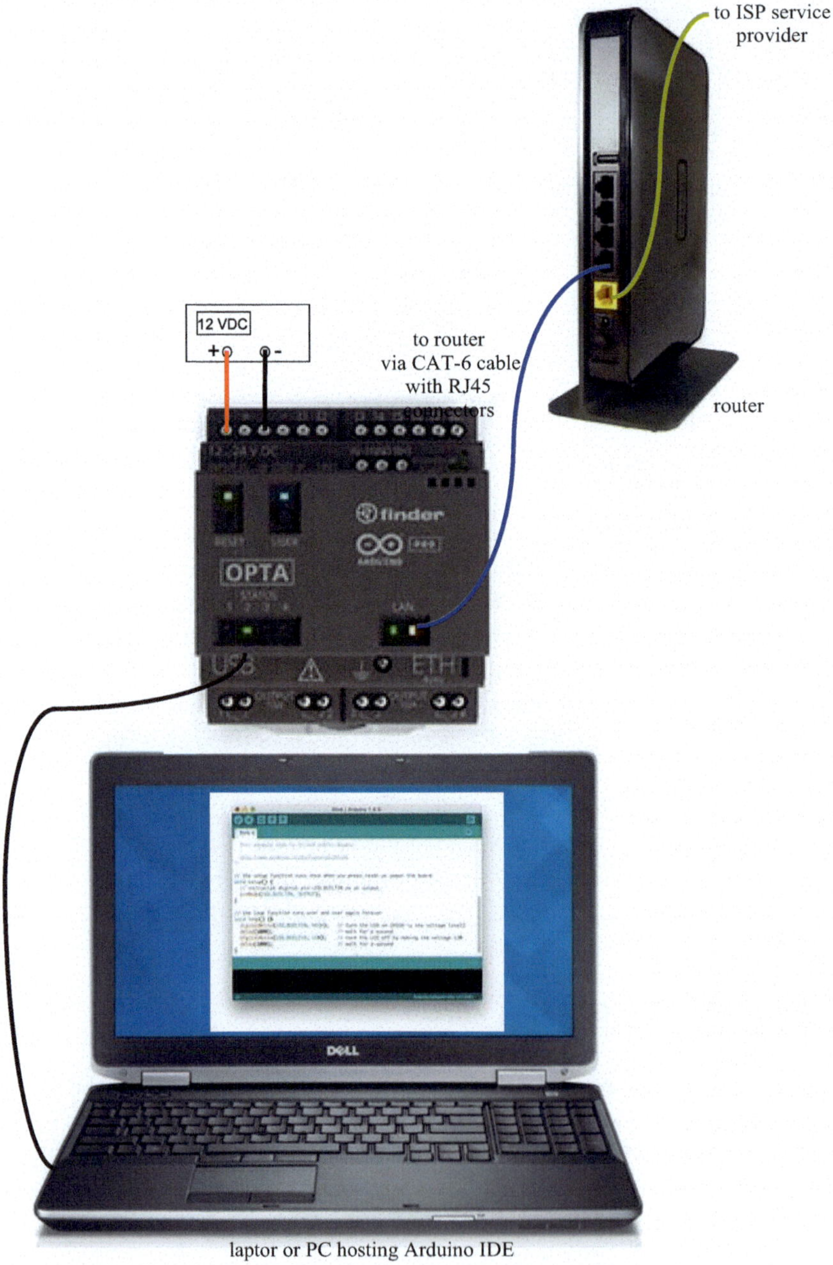

Fig. 2.20 Ethernet configuration. Images used courtesy of the Arduino team (CC BY–NC–SA) (www.arduino.cc)

(API). When queried the API provides the geolocation of the IP address (www.ip-api.com) and provides the result on the Serial Monitor.

Note the sketch's use of the Arduino Ethernet and JSON header files. These are available for download via the Library Manager within the Arduino IDE. JSON or Java Script Object Notation is a standard for data interchange in a human readable format (www.json.org).

```
//**********************************************************************
//Web Client (Ethernet version)
//Name: opta_ethernet_web_client.ino
//Purpose: This sketch connects an Opta device to ip-api.com via Ethernet
//and fetches IP details for the device.
//
//@author Arduino PRO Content Team
//@version 2.0 15/08/23
//**********************************************************************

#include <Ethernet.h>                        //Include libraries
#include <Arduino_JSON.h>

const char* server = "ip-api.com";           //Server addr ip-api.com

String path = "/json/";                      //API path for IP details

IPAddress ip(10, 130, 22, 84);               //Static IP config for Opta

//Ethernet client instance for the communication
EthernetClient client;

//JSON variable to store and process the fetched data
JSONVar doc;

//Variable to ensure we fetch data only once
bool dataFetched = false;

void setup()
{
Serial.begin(115200);                        //serial comm at 115200 Baud
while (!Serial);                             //wait for serial port connection

//Attempt to start Ethernet connection via DHCP.
//If DHCP fails, print diagnostic message.
  if(Ethernet.begin() == 0)
    {
    Serial.println("- Failed to configure Ethernet using DHCP!");

    //Try to configure Ethernet with the predefined static IP address.
    Ethernet.begin(ip);
  }
  delay(2000);
}

void loop()
{
//Ensure we haven't fetched data already, ensure the Ethernet link is
//active, establish a connection to the server, and compose and send the
```

```
// HTTP GET request.
if(!dataFetched)
  {
  if(Ethernet.linkStatus() == LinkON)
    {
    if(client.connect(server, 80))
      {
      client.print("GET ");
      client.print(path);
      client.println(" HTTP/1.1");
      client.print("Host: ");
      client.println(server);
      client.println("Connection: close");
      client.println();

      //Wait and skip the HTTP headers to get to the JSON data.
      char endOfHeaders[] = "\r\n\r\n";
      client.find(endOfHeaders);

      //Read and parse the JSON response.
      String payload = client.readString();
      doc = JSON.parse(payload);

      //Check if the parsing was successful.
      if(JSON.typeof(doc) == "undefined")
        {
        Serial.println("- Parsing failed!");
        return;
        }

      //Extract and print the IP details.
      Serial.println("*** IP Details:");
      Serial.print("- IP Address: ");
      Serial.println((const char*)doc["query"]);
      Serial.print("- City: ");
      Serial.println((const char*)doc["city"]);
      Serial.print("- Region: ");
      Serial.println((const char*)doc["regionName"]);
      Serial.print("- Country: ");
      Serial.println((const char*)doc["country"]);
      Serial.println("");

      //Mark data as fetched.
      dataFetched = true;
      }
    //Close the client connection once done.
    client.stop();
  }
  else
    {
    Serial.println("- Ethernet link disconnected!");
    }
  }
}

//**********************************************************************
```

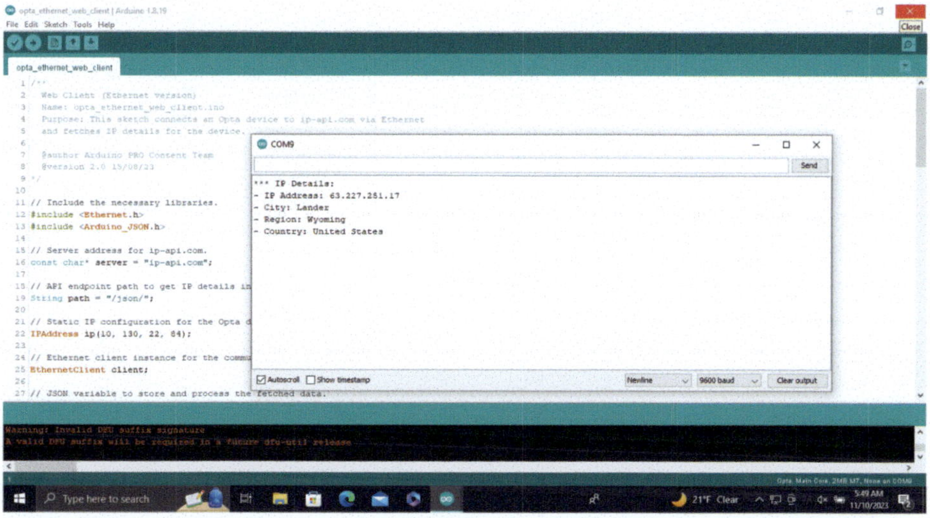

Fig. 2.21 Ethernet sketch result

The result is shown in Fig. 2.21.

Example: Ethernet 2. The next example, Ethernet WebClient, is available within the Arduino Mbed OS Opta Boards core. Additional instructions are provided within the "Bluetooth Low Energy, Wi–Fi and Ethernet with OPTA (www.opta.findernet.com/en/)." When compiled, uploaded, and executed; the sketch goes to the designated website and retrieves and displays the website's contents to the serial monitor.

```
//*********************************************************************
//WebClient. This sketch connects to a website (http://www.google.com)
//and displays results to the Serial Monitor.
//
//created 18 Dec 2009 by David A. Mellis
//modified 9 Apr 2012 by Tom Igoe based on work by Adrian McEwen
//*********************************************************************

#include <PortentaEthernet.h>
#include <Ethernet.h>
#include <SPI.h>

//Enter a MAC address for your controller below.
//byte mac[] = { 0xDE, 0xAD, 0xBE, 0xEF, 0xFE, 0xED };
// if you don't want to use DNS (and reduce your sketch size)
// use the numeric IP instead of the name for the server:
//IPAddress server(74,125,232,128);    //numeric IP for Google (no DNS)
char server[] = "www.google.com";      //name addr for Google (using DNS)

//Set the static IP address to use if the DHCP fails to assign
```

```
IPAddress ip(192, 168, 2, 177);
IPAddress myDns(192, 168, 2, 1);

//Initialize the Ethernet client library with the IP address and
//port of the server that you want to connect to (port 80 is default
//for HTTP):
EthernetClient client;

//Variables to measure the speed
unsigned long beginMicros, endMicros;
unsigned long byteCount = 0;
bool printWebData = true;              //set false-better speed measure

void setup()
{
//Open serial communications and wait for port to open:
Serial.begin(9600);
while (!Serial)
  {
  ;// wait for serial port to connect. Needed for native USB port only
  }

//start the Ethernet connection:
Serial.println("Initialize Ethernet with DHCP:");
if(Ethernet.begin() == 0)
  {
  Serial.println("Failed to configure Ethernet using DHCP");
  //Check for Ethernet hardware present
  if(Ethernet.hardwareStatus() == EthernetNoHardware)
    {
    Serial.println("Ethernet shield was not found.Ã,âŁž);
    Serial.println("Sorry, can't run   without hardware. :(");
    while(true)
      {
      delay(1); //do nothing, no point running without Ethernet hardware
      }
    }
    if(Ethernet.linkStatus() == LinkOFF)
      {
      Serial.println("Ethernet cable is not connected.");
      }
    //try to congifure using IP address instead of DHCP:
    Ethernet.begin(ip, myDns);
    }
    else
      {
      Serial.print("  DHCP assigned IP ");
      Serial.println(Ethernet.localIP());
      }
 //give the Ethernet shield a second to initialize:
 delay(1000);
 Serial.print("connecting to ");
 Serial.print(server);
```

```
 Serial.println("...");

 //if you get a connection, report back via serial:
 if(client.connect(server, 80))
   {
   Serial.print("connected to ");
   Serial.println(client.remoteIP());

   //Make a HTTP request:
   client.println("GET /search?q=arduino HTTP/1.1");
   client.println("Host: www.google.com");
   client.println("Connection: close");
   client.println()
   }
 else
   {
   //if you didn't get a connection to the server:
   Serial.println("connection failed");
   }
beginMicros = micros();
}

void loop()
{
//if there are incoming bytes available from the server,
//read them and print them:
int len = client.available();
if(len > 0)
  {
  byte buffer[80];
  if(len > 80)
     len = 80;
  client.read(buffer, len);
  if(printWebData)
    {
    Serial.write(buffer, len);          //show serial monitor
    }
        byteCount = byteCount + len;
  }
//if the server's disconnected, stop the client:
if(!client.connected())
  {
  endMicros = micros();
  Serial.println();
  Serial.println("disconnecting.");
  client.stop();
  Serial.print("Received ");
  Serial.print(byteCount);
  Serial.print(" bytes in ");
  float seconds = (float)(endMicros - beginMicros) / 1000000.0;
  Serial.print(seconds, 4);
  float rate = (float)byteCount / seconds / 1000.0;
  Serial.print(", rate = ");
```

```
    Serial.print(rate);
    Serial.print(" kbytes/second");
    Serial.println();

    //do nothing forevermore:
    while(true)
      {
      delay(1);
      }
    }
}

//**********************************************************************
```

2.3 Variant Specific Features

In this section we explore Opta PLC features available on specific variants. For each feature we identify variants hosting the feature, some basic theory of operation, and an illustrative example.

2.3.1 Wi–Fi 802.11 B/g/n (Opta WiFi)

The Opta WiFi variant (AFX000002) is equipped with hardware to support the 802.11 b/g/n Wireless Fidelity (Wi–Fi) standard. The b/g/n suffix designates different bit rate and frequency features for 802.11 versions: b (11 Mbps, 2.4 GHz, 1999); g(6–54 Mbps, 2.4 GHz, 2003); and n (72–600 Mbps, 2.4 and 5 Ghz, 2008). Opta Wi–Fi features also includes support for WEP, WPA, WPA2, and WPA3 security standards. The Opta also hosts an internal Wi–Fi antenna.

The next two examples are Wi–Fi versions of the two previous ethernet examples.

Example: Wi–Fi 1. For this demonstration, an Ardunio Opta is connected to a home router via Wi–Fi. The Opta is programmed using a laptop/PC. In the sketch the Opta is configured as a client. It will connect to a server configured computer via the internet and request the service provided. In this specific example the server is located at www.ip-api.com. This website hosts an IP Geolocation Application Programming Interface (API). When queried the API provides the geolocation of the IP address (www.ip-api.com) and provides the result on the Serial Monitor.

Note the sketch's use of the Arduino Wi–Fi and JSON header files. These are available for download via the Library Manager within the Arduino IDE. JSON or Java Script Object Notation is a standard for data interchange in a human readable format (www.json.org).

```
//************************************************************************
//WiFi Web Client
//Name: opta_wifi_web_client.ino
//Purpose: This sketch connects an Opta device to ip-api.com via WiFi
//and fetches IP details.
//
//@author Arduino PRO Content Team
//@version 2.2 16/08/23
//************************************************************************

#include <WiFi.h>
#include <Arduino_JSON.h>

//Wi-Fi network details.
const char* ssid     = "insert your network name";
const char* password = "insert your network passwor";

//Server address for ip-api.com.
const char* server = "ip-api.com";

//API endpoint path to get IP details in JSON format.
String path = "/json";

//Wi-Fi client instance for the communication.
WiFiClient client;

//JSON variable to store and process the fetched data.
JSONVar doc;

//Variable to ensure we fetch data only once.
bool dataFetched = false;

void setup()
{
//Begin serial communication at a baud rate of 115200.
Serial.begin(115200);

//Wait for the serial port to connect,
//This is necessary for boards that have native USB.
while (!Serial);

//Start the Wi-Fi connection using the provided SSID and password.
Serial.print("- Connecting to ");
Serial.println(ssid);
WiFi.begin(ssid, password);

while(WiFi.status() != WL_CONNECTED)
  {
  delay(1000);
  Serial.print(".");
  }

Serial.println();
```

```
Serial.println("- Wi-Fi connected!");
Serial.print("- IP address: ");
Serial.println(WiFi.localIP());
}

void loop()
{
//Check if the IP details have been fetched.
//If not, call the function to fetch IP details,
//Set the flag to true after fetching.
if(!dataFetched)
  {
  fetchIPDetails();
  dataFetched = true;
  }
}

//**********************************************************************
//Fetch IP details from defined server
//@param none
//@return IP details
//**********************************************************************

void fetchIPDetails()
  {
  if(client.connect(server, 80))
    {
    //Compose and send the HTTP GET request.
    client.print("GET ");
    client.print(path);
    client.println(" HTTP/1.1");
    client.print("Host: ");
    client.println(server);
    client.println("Connection: close");
    client.println();

    //Wait and skip the HTTP headers to get to the JSON data.
    char endOfHeaders[] = "\r\n\r\n";
    client.find(endOfHeaders);

    //Read and parse the JSON response.
    String payload = client.readStringUntil('\n');
    doc = JSON.parse(payload);

    //Check if the parsing was successful.
    if(JSON.typeof(doc) == "undefined")
      {
      Serial.println("- Parsing failed!");
      return;
      }

    //Extract and print the IP details.
    Serial.println("*** IP Details:");
```

```
    String query = doc["query"];
    Serial.print("- IP Address: ");
    Serial.println(query);
    String city = doc["city"];
    Serial.print("- City: ");
    Serial.println(city);
    String region = doc["regionName"];
    Serial.print("- Region: ");
    Serial.println(region);
    String country = doc["country"];
    Serial.print("- Country: ");
    Serial.println(country);
    Serial.println("");
    }
  else
    {
    Serial.println("- Failed to connect to server!");
    }

//Close the client connection once done.
client.stop();
}

//***********************************************************************
```

Example: Opta Wi–Fi 2. The next example, WiFi WebClient, is available within the Arduino Mbed OS Opta Boards core. Additional instructions are provided within the "Bluetooth Low Energy, Wi–Fi and Ethernet with OPTA (www.opta.findernet.com/en/)." When compiled, uploaded, and executed; the sketch goes to the designated website and retrieves and displays the website's contents to the serial monitor.

```
//***********************************************************************
//Web client: This sketch connects to a website (http://example.com)
//using the WiFi module.  This example is written for a network using
//WPA encryption. For WEP or WPA, change the Wifi.begin() call
//accordingly.
//
//created 13 July 2010 by dlf (Metodo2 srl)
//modified 31 May 2012 by Tom Igoe
//***********************************************************************


#include <WiFi.h>

#include "arduino_secrets.h"
//enter your sensitive data in the Secret tab/arduino_secrets.h
char ssid[] = SECRET_SSID;        //your network SSID (name)
char pass[] = SECRET_PASS;        //your network password (use for WPA,
```

```
                                        //or use as key for WEP)
int keyIndex = 0;                       //your network key Index number
                                        //(needed only for WEP)

int status = WL_IDLE_STATUS;
//if you don't want to use DNS (and reduce your sketch size)
//use the numeric IP instead of the name for the server:
//IPAddress server(93,184,216,34);//IP address for example.com (no DNS)
char server[] = "example.com";     //host name for example.com (using DNS)

WiFiClient client;

void setup()
{
//Initialize serial and wait for port to open:
Serial.begin(9600);
while (!Serial)
  {
  ; //wait for serial port to connect. Needed for native USB port only
  }

//check for the WiFi module:
if(WiFi.status() == WL_NO_SHIELD)
  {
  Serial.println("Communication with WiFi module failed!");
  //don't continue
  while (true);
  }

//attempt to connect to Wifi network:
while(status != WL_CONNECTED)
  {
  Serial.print("Attempting to connect to SSID: ");
  Serial.println(ssid);
  //Connect to WPA/WPA2 network. Change line if using open or WEP
  //network:
  status = WiFi.begin(ssid, pass);

  //wait 3 seconds for connection:
  delay(3000);
  }

Serial.println("Connected to wifi");
printWifiStatus();

Serial.println("\nStarting connection to server...");
//if you get a connection, report back via serial:
if(client.connect(server, 80))
  {
  Serial.println("connected to server");

  //Make a HTTP request:
  client.println("GET /index.html HTTP/1.1");
```

```
  client.print("Host: ");
  client.println(server);
  client.println("Connection: close");
  client.println();
  }
}

void loop()
{
//if there are incoming bytes available
//from the server, read them and print them:
while (client.available())
  {
  char c = client.read();
  Serial.write(c);
  }

//if the server's disconnected, stop the client:
if(!client.connected())
  {
  Serial.println();
  Serial.println("disconnecting from server.");
  client.stop();

  //do nothing forevermore:
  while (true);
  }
}

//*********************************************************************
void printWifiStatus()
{
//print the SSID of the network you're attached to:
Serial.print("SSID: ");
Serial.println(WiFi.SSID());

//print your board's IP address:
IPAddress ip = WiFi.localIP();
Serial.print("IP Address: ");
Serial.println(ip);

//print the received signal strength:
long rssi = WiFi.RSSI();
Serial.print("signal strength (RSSI):");
Serial.print(rssi);
Serial.println(" dBm");
}

//*********************************************************************
```

2.3.2 RS–485 Communication (Opta WiFi and Opta RS485)

The RS–485 (TIA/EIA–485) is an electrical standard providing for communication between devices in a noisy, industrial environment. The standard does not define a specific communication protocol but provides the electrical specifications to implement specific protocols such as MODBUS RTU.

The RS–485 provides for one–way (half–duplex) communication between industrial devices using a differential signaling scheme. As shown in Fig. 2.22, there are two signal lines designated $A(-)$ and $B(+)$ and also a shared ground connection. A logic one (Mark) is signaled with a low signal on the $A(-)$ terminal and a high signal on the $B(+)$ terminal. A logic low (Space) is signaled with a high signal on the $A(-)$ terminal and a low signal on the $B(+)$ terminal. A twisted pair cable is used to connect RS–485 compatible devices.

The RS–485 standard provides for cable runs up to 1200 m, at data rates up to 10 M bits/second, and up to 32 RS–485 configured units. There is a tradeoff between maximum data rate and cable length. Depending on the specific application, system requirements, and configuration; termination resistors may be required at either end of the RS–485 cable run. In general, long cable runs transmitting at high data rates require termination resistors. Typically a 120 Ω resistor is used for termination (Horowitz and Hill).

Example. In this example we connect two Opta PLCs equipped with RS—485 via a 24 AWG twisted pair cable and common ground. See Fig. 2.22. One PLC is designated and equipped as the sender and the other as a receiver. For successful communication between the sender and receiver both must be configured with the same data Baud rate. Although the two Opta are shown near one another in the diagram, in an industrial setting there may be some distance between the two units.[5]

The overall goal of the example is to link the sender and receiver via an RS–485 link. At the sender PLC, the user will insert a number between one and four to designate which relay to toggle on the receiver PLC. We have connected a red 10 mm LED to each relay output. A Unified Modeling Diagram (UML) activity diagram is provided in Fig. 2.23 to illustrate the operation of the sender and receiver sketch.

Aside–UML activity diagram: The activity diagram is simply a UML compliant flow chart. UML is a standardized method of documenting systems. The activity diagram is one of the many tools available from UML to document system design and operation. The basic symbols used in a UML activity diagram for a processor based system are provided in Fig. 2.24 (Fowler).

[5] Both send and receive sketches are available from " Getting Started with RS–485 on Opta, www.opta.findernet.com.

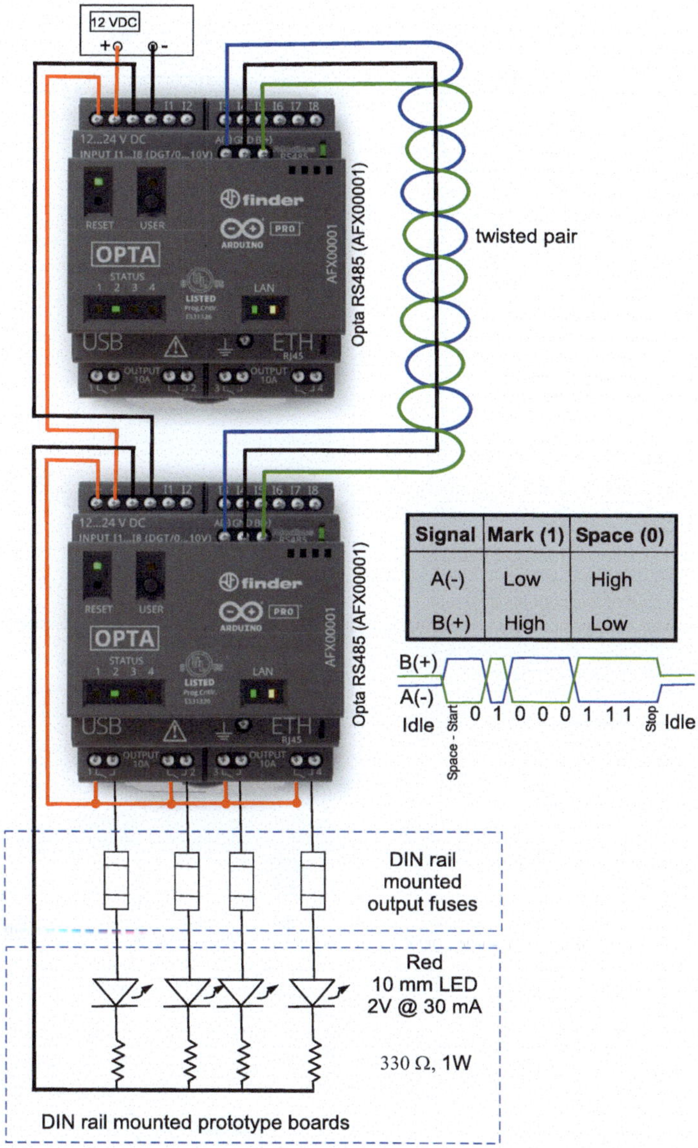

Fig. 2.22 Arduino Opta RS–485. Images used courtesy of the Arduino team (CC BY–NC–SA) (www.arduino.cc)

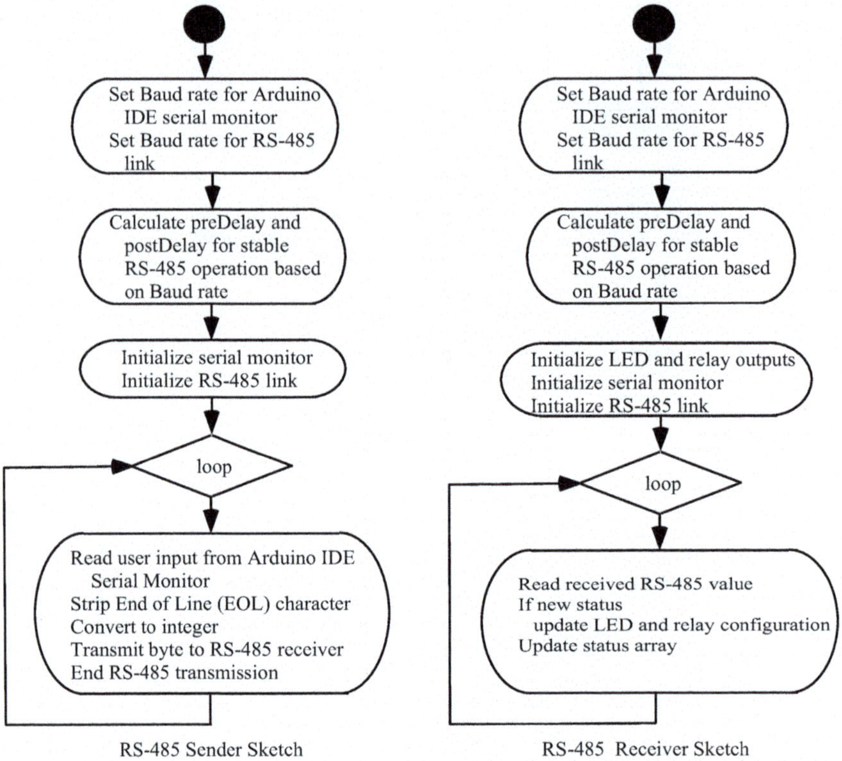

Fig. 2.23 Arduino opta RS–485

Fig. 2.24 UML activity diagram symbols. Adapted from (Fowler)

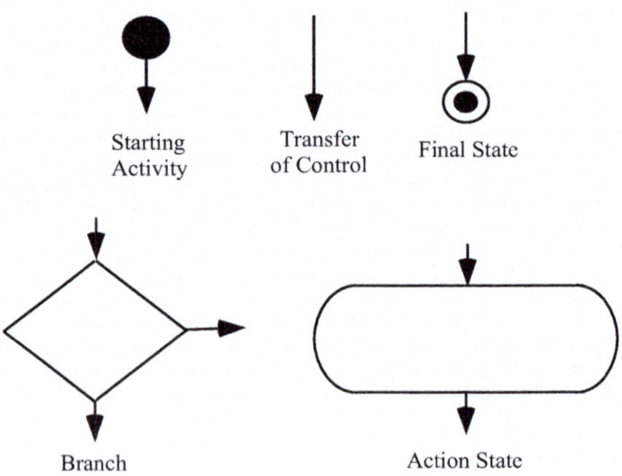

Opta RS485 Sender sketch

```
//*******************************************************************
//Opta_RS485_Sender sketch
//Source: Getting Started with RS-485 on OptaÃ‚â„¢
//Name: Opta_RS485_Sender
//Purpose: Sends values input by user via the Serial monitor
//to Receiver Opta.
//
//author Giampaolo Mancini, Arduino
//@version 1.0 02/02/2023
//*******************************************************************

#include <ArduinoRS485.h>

constexpr auto baudrate{115200};

//Calculate preDelay and postDelay in microseconds for stable
//RS-485 transmission
constexpr auto bitduration{1.f / baudrate};
constexpr auto wordlen{9.6f};        //or 10.0f depending on ch config
constexpr auto preDelayBR{ bitduration * wordlen * 3.5f * 1e6 };
constexpr auto postDelayBR{ bitduration * wordlen * 3.5f * 1e6 };

void setup()
{
Serial.begin(baudrate);
while (!Serial);

RS485.begin(baudrate);
RS485.setDelays(preDelayBR, postDelayBR);
}

void loop()
{
auto aval = Serial.available();
if(aval > 0)
  {
  auto input = Serial.readStringUntil('\r');
  auto read = input.length();                  //discard EOL
  while (aval > ++read)
  Serial.read();
  auto incomingByte = input.toInt();
  RS485.beginTransmission();
  Serial.print("- Sending: ");
  Serial.println(incomingByte);
  RS485.write(incomingByte);
  RS485.endTransmission();
  }
}

//*******************************************************************
```

Opta RS485 Receiver sketch

```
//*****************************************************************
//Opta_RS485_Receiver sketch
//Source: Getting Started with RS-485 on OptaÃ,â„¢
//Name: Opta_RS485_Receiver
//Purpose: Captures values from Sender Opta and triggers
//corresponding relay and status LED.
//
//@author Giampaolo Mancini, Arduino
//@version 1.0 02/02/2023
//*****************************************************************

#include <ArduinoRS485.h>

constexpr auto baudrate{115200};

//Calculate preDelay and postDelay in microseconds for stable RS-485
//transmission constexpr auto bitduration{1.f / baudrate};
constexpr auto wordlen{ 9.6f };     //or 10.0f depending on ch config
constexpr auto preDelayBR{ bitduration * wordlen * 3.5f * 1e6 };
constexpr auto postDelayBR{ bitduration * wordlen * 3.5f * 1e6 };

int idx{0};
bool newState{false};
int relays[]{D0, D1, D2, D3};
int leds[]{LED_D0, LED_D1, LED_D2, LED_D3};
bool statuses[]{ true, true, true, true };

void setup()
{
for(int i = 0; i < 4; i++)
  {
  pinMode(relays[i], OUTPUT);
  pinMode(leds[i], OUTPUT);
  }

RS485.begin(baudrate);
RS485.setDelays(preDelayBR, postDelayBR);
Serial.begin(baudrate);
//while (!Serial);
}

void loop()
{
RS485.receive();
auto aval = RS485.available();
if(aval > 0)
  {
  int readValue = RS485.read();
  if(readValue > 4)                     //manage out-of-range inputs
  readValue = readValue % 4;
  Serial.print("Command for relay: ");
  Serial.println(readValue);
  newState = true;
  idx = readValue - 1;                  //array indexes start at 0
  }
```

```
  RS485.noReceive();
  if(newState)
    {
    changeRelay();
    newState = false;
    }
  }

//********************************************************************
//void changeRelay()- changes relay and status LED state given the
//received value.
//********************************************************************

void changeRelay()
{
auto status = statuses[idx] ? HIGH:LOW; //get current status
digitalWrite(relays[idx], status);      //apply new status to outputs
digitalWrite(leds[idx], status);
statuses[idx] = !statuses[idx];         //invert status array for update
}

//********************************************************************
```

Note in the Opta RS485 Receiver sketch, we have commented out the "while(!Serial);" line of the code. If left in the sketch, the program waits for the Serial Monitor to be opened by the user. We load the receiver sketch to the receiver designated PLC first. We then load the sender sketch to the sender designated PLC. The Serial Monitor on the sender designated PLC is opened to insert numbers between one and four to toggle relay status on the receiver PLC. The completed configuration is shown in Fig. 2.25.

2.3.3 Bluetooth Low Energy (BLE) (Opta WiFi)

The Arduino Opta WiFi is equipped with Bluetooth features. The Classic form of Bluetooth was designed to provide a wireless replacement for the common RS–232 serial connection standard. The Arduino Opta WiFi is also equipped with Bluetooth Low Energy (BLE) features. It is important to note that Bluetooth Classic and BLE features are not compatible with one another.[6]

Bluetooth BLE provides for low transmit power (10 mW), short (maximum 100 m) range RF connections to replace wires. It uses the crowded Industrial, Scientific, and Medical (ISM) frequency band from 2.40 to approximately 2.50 GHz. The BLE band is divided into 40 different, 2 MHz channels as shown in Fig. 2.26. BT BLE employs an interesting frequency hopping technique to communicate. Data for transmission is divided into packets at data rates from 125 to 2 Mb/s. The device transmits a packet of data at the first carrier frequency. It then hops to a different carrier frequency for the next packet and so on until

[6] Portions of this section are provided with permission from "Arduino V: Machine Learning," S. Barrett, Springer, 2022.

Fig. 2.25 Arduino opta RS–485 panel

the entire message is transmitted as shown in Fig. 2.26b. Formally the BT BLE modulation technique is called Direct Sequence Spread Spectrum (DSSS) (www.bluetooth.com).

BLE uses the Generic Attribute (GATT) Profile to establish two different primary roles for a BLE connection:

- The peripheral or server role provides bulletin board features where data is posted for reading.
- The central or client role can read and interact with the posted data.

In Fig. 2.27 we use an Arduino Opta WiFi in a peripheral server role to collect important greenhouse information such as external temperature, internal temperature, humidity, and soil moisture content. The greenhouse related data is collected and organized into a BLE service. The service related data is provided as BLE configured characteristics. To allow ease of access to the information from an external central client device, the BLE service and characteristics are each assigned a universally unique identifier (UUID) (www.bluetooth.

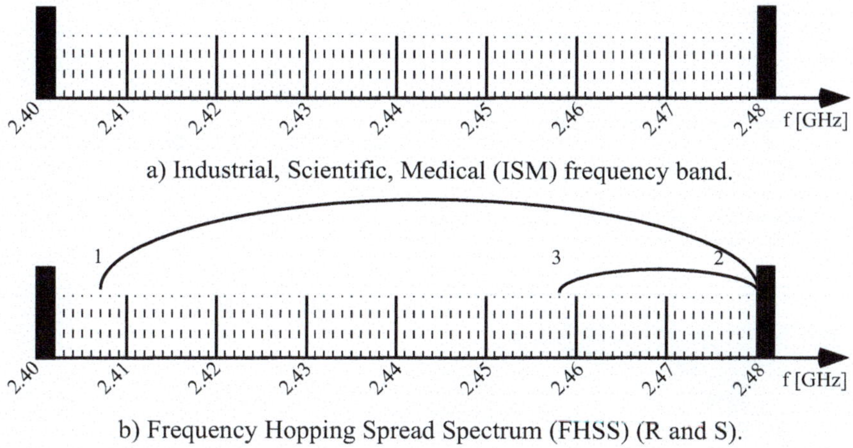

Fig. 2.26 Bluetooth BLE communication concepts

com). If we were to expand the features of the project with additional services, we could group them into a profile.

There are a number of 16 bit pre–assigned UUIDs. The UUIDs represent different manufacturers and technology companies employing Bluetooth–based technologies. Also, UUIDs have been pre–assigned to common Bluetooth features and common pre–assigned data types (e.g. temperature, pressure, etc.) (www.bluetooth.com):

- Bluetooth members: 0xFxxx
- GATT characteristic and object type: 0x2xxx
- GATT declarations: 0x28xx and 0x29xx
- GATT service: 0x18xx
- GATT unit: 0x27xx
- protocol identifier: 0x00xx
- SDO GATT service: 0xFFFx
- service classes and profiles: 0x10xx and 0x11xx.

For BLE services and characteristics without a 16 bit pre–assigned UUID, a unique 128 bit UUID code is used. A Bluetooth unique UUID may be obtained using a number of online UUID generators.

In the greenhouse example, a cell phone is configured as a BLE central or client. Through the BLE wireless radio interconnect, the cell phone can read and interact with the greenhouse data and features.

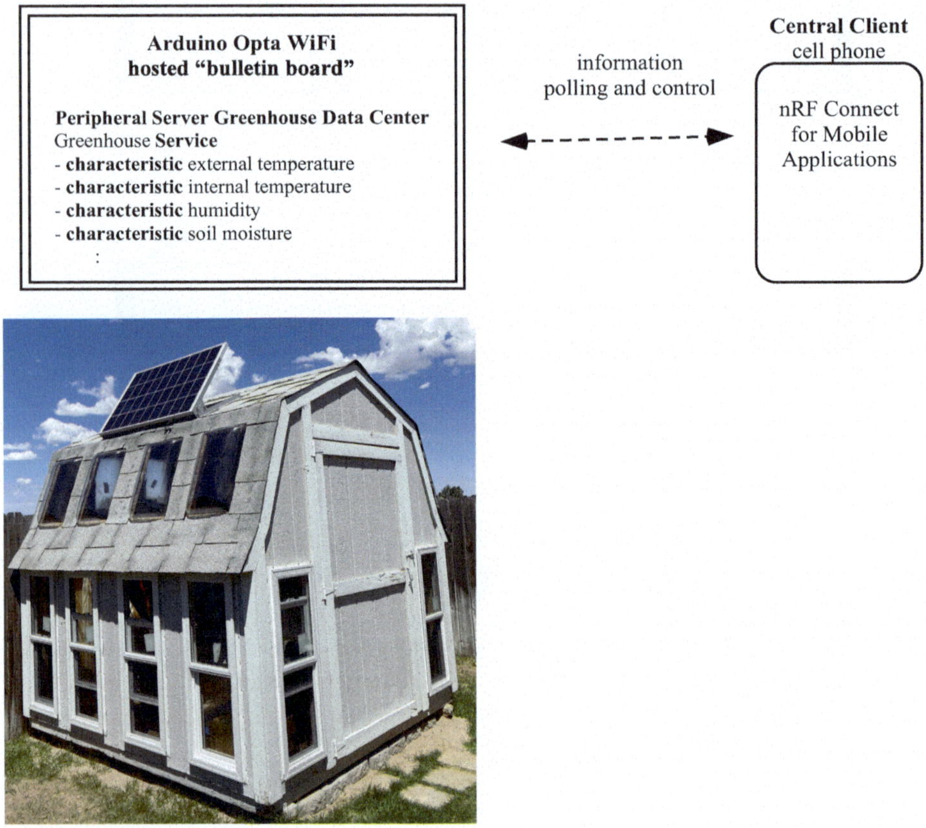

Fig. 2.27 Bluetooth BLE equipped greenhouse

2.3.3.1 ArduinoBLE Library

The ArduinoBLE Library provides for a wide variety of BLE configurations. The library is downloaded from within the Arduino IDE using the Library Manager. The library is organized into different classes including the (www.arduino.cc):

- BLE Class used to enable the BLE module,
- BLE Device Class to get information about connected devices,
- BLE Service Class to enable services and interaction with services,
- BLE Characteristic Class to enable characteristics and interaction with them, and
- BLE Descriptor Class to describe characteristics.

To get acquainted with the library we continue with a series of examples. The first two examples are adapted from the Arduino BLE Library. In the third example, we configure an Arduino Opta WiFi as the server to collect and post greenhouse data. A cell phone is configured as a client to poll and interact with the greenhouse data. The cell phone is equipped with a BLE compatible app to interact with the Opta WiFi. This example is provided in the Application section at the end of the chapter.

Example: In this first example "LED," from the Arduino BLE Library, a cell phone serves as a central client to control an LED onboard the Nano 33 BLE Sense configured as a server.

To get better acquainted with the sketch, we study the Bluetooth configuration related code steps. In Fig. 2.28 we detail these steps in a UML activity diagram.

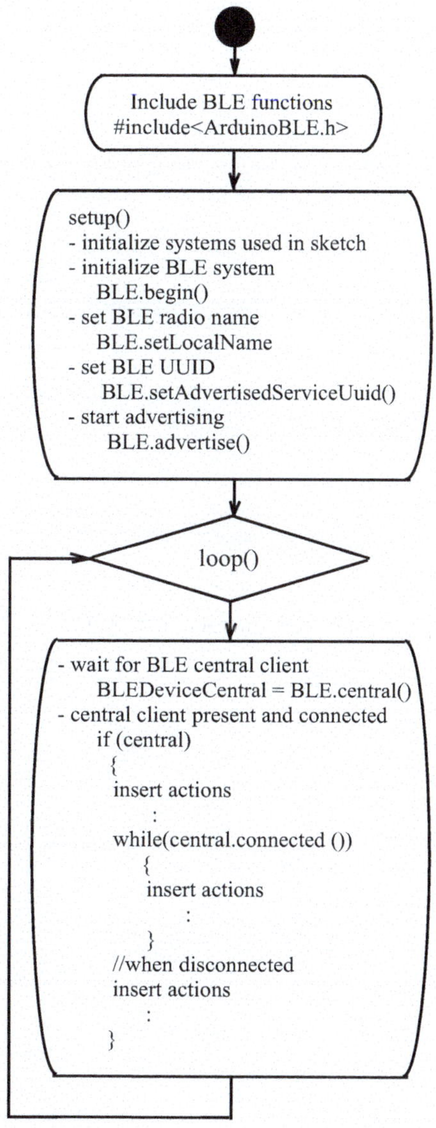

Fig. 2.28 Bluetooth BLE configuration

```
//**************************************************************
//Opta_BLE_LED: This example creates a BLE peripheral with
//service that contains a characteristic to control an LED.
//
//A generic BLE central phone app, like LightBlue or
//nRF Connect is used to interact with the Arduino Opta
//WiFi (AFX00002) hosted BLE services and characteristics
//created in this sketch.
//
//This example code is in the public domain.
//***************************************************************

#include <ArduinoBLE.h>
                                     //Declare BLE LED Service
                                     //Link to 128 bit UUID
BLEService ledService("19B10000-E8F2-537E-4F6C-D104768A1214");

//BLE LED Switch Characteristic - custom 128-bit UUID, read and
//writable by central client device (cell phone)
BLEByteCharacteristic switchCharacteristic
   ("19B10001-E8F2-537E-4F6C-D104768A1214", BLERead | BLEWrite);

const int ledPin = LED_BUILTIN;      //Use builtin LED

void setup()
{
Serial.begin(9600);                  //status to serial monitor
while (!Serial);

pinMode(ledPin, OUTPUT);             //set LED pin to output mode

if(!BLE.begin())                     //BLE initialization
  {
  Serial.println("starting BLE failed!");
  while (1);
  }

  //set advertised local name and service UUID:
  BLE.setLocalName("LED");
  BLE.setAdvertisedService(ledService);

  //add the characteristic to the service
  ledService.addCharacteristic(switchCharacteristic);

  //add service
  BLE.addService(ledService);

  //set the initial value for the characeristic:
  switchCharacteristic.writeValue(0);

  //start advertising
```

```
  BLE.advertise();

  Serial.println("BLE LED Peripheral");
}

void loop()
{
//listen for BLE clients (central) to connect:
BLEDevice central = BLE.central();

//if a client (central) is connected to peripheral:
if(central)
  {
  Serial.print("Connected to client: ");

  //print the client's MAC address:
  Serial.println(central.address());

  //while the client (central) is still connected to
  //the Opta WiFi based server (peripheral):
  while (central.connected())
    {
    //if the remote client device wrote to the
    //Opta WiFi server characteristic, use the
    //value to control the LED:
      if(switchCharacteristic.written())
        {
        if(switchCharacteristic.value())
          {
          Serial.println("LED on");      //any value other than 0
          digitalWrite(ledPin, HIGH);    //will turn the LED on
          }
        else
          {
          Serial.println(F("LED off")); //a 0 value
          digitalWrite(ledPin, LOW);     //will turn the LED off
          }
        }
     }//end while

  //when the central disconnects, print it out:
  Serial.print(F("Disconnected from central: "));
  Serial.println(central.address());
  }//end if(central)
}

//***************************************************************
```

The sketch may be compiled and uploaded to the Arduino Opta WiFi. Once uploaded, the sketch may be tested:

- Open the Serial Monitor in the Arduino IDE to monitor sketch status.
- Using a cell phone as a client, open "nRF Connect" to establish Bluetooth BLE connection with the Opta WiFi based server.[7]
- Find "LED" in the nRF scanner list.
- Tap "Connect" to connect the client (cell phone) to the server (Arduino Opta).
- By selecting "Client" and the up arrow, values may be sent from the client to the server to control the LED.
- Select "Write Value" and "Unsigned."
- Sending a non–zero turns the LED on while sending zero turns the LED off.

Example: In this example, "battery monitor," adapted from the Arduino BLE Library, the Arduino Opta WiFi is configured as a server. The Arduino Opta WiFi monitors the analog signal on Input I1 (A0) and posts this characteristic to the server based bulletin board. A cell phone based client equipped with BLE compatible app is used to poll the posted data.

To simulate a battery, a 1M Ohm potentiometer is connected to pin A0. The potentiometer is connected between 10 VDC and ground. The potentiometer wiper arm (center terminal) is connected to the Input I1 (A0) pin.

The client/server connection is tested using techniques similar to those provided in the previous example.

```
//**********************************************************
//Opta_battery_monitor - This example creates a BLE server
//(peripheral) with the standard battery service and level
//characteristic. The Input I1 (A0) pin is used to monitor the
//battery level.
//
//A generic BLE central phone app, like LightBlue or
//nRF Connect is used to interact with the Arduino Opta WiFi
//hosted BLE services and characteristics created in this
//sketch.
//
//This example code is in the public domain.
//**********************************************************

#include <ArduinoBLE.h>

BLEService batteryService("180F");            //BLE Battery Service

//BLE Battery Level Characteristic
//  - standard 16-bit characteristic UUID
//  - remote clients get notifications if characteristic changes
BLEUnsignedCharCharacteristic batteryLevelChar("2A19",
                                           BLERead | BLENotify);
int battery;                                  //battery reading
int oldBatteryLevel = 0;                      //last battery level reading from A0
long previousMillis = 0;                      //last time battery level checked (ms)

void setup()
```

[7] BLE applications such as nRF connect or LightBlue are available from your cell phone app store.

```
{
Serial.begin(9600);                          //initialize serial communication
while (!Serial);

                                             //init built-in LED pin
pinMode(LED_BUILTIN,OUTPUT);                 //indicates when central is connected
analogReadResolution(12);                    //set 12 to 16 bits


if(!BLE.begin())                             //initialize Bluetooth BLE device
  {
  Serial.println("starting BLE failed!");
  while (1);
  }

//Set a local name for the BLE device. This name appears
//in advertising packets. Name used by remote devices to
//identify this BLE device.
BLE.setLocalName("BatteryMonitor");
BLE.setAdvertisedService(batteryService);     //add the service UUID
                                              //add the battery level characteristic
batteryService.addCharacteristic(batteryLevelChar);
BLE.addService(batteryService);               //add battery service
batteryLevelChar.writeValue(oldBatteryLevel); //set initial value

//Start advertising BLE.  Continuously transmits BLE advertising
//packets. Advertising will be visible to remote BLE central devices.
BLE.advertise();
Serial.println("Bluetooth device active, waiting for connections...");
}

void loop()
{
BLEDevice central = BLE.central();            //wait for a BLE central
                                              //if a central client
                                              //is connected to peripheral
if(central)
{
Serial.print("Connected to central: ");
Serial.println(central.address());            //print the central's BT address
digitalWrite(LED_BUILTIN, HIGH);              //LED on when client connected
                                              //while client connected
                                              //check battery level every 200ms
while(central.connected())
  {
  long currentMillis = millis();
                                              //if 200ms have passed,
                                              //check the battery level
  if(currentMillis - previousMillis >= 200)
    {
    previousMillis = currentMillis;
    updateBatteryLevel();
    }
  }

  //when the central client disconnects, turn off the LED
  digitalWrite(LED_BUILTIN, LOW);
  Serial.print("Disconnected from central: ");
  Serial.println(central.address());
  }
```

```
}

//************************************************************
//void updateBatteryLevel() - Read the current voltage level
//on the A0 analog input pin. This is used here to simulate
//the battery level.
//************************************************************

void updateBatteryLevel()
{
//Read the input on analog input I1 corresponding to A0:
battery = analogRead(A0);

int batteryLevel = map(battery, 0, 4095, 0, 100);

if(batteryLevel != oldBatteryLevel)
  {                                    //if the battery level has changed
  Serial.print("Battery Level %: ");   // print it
  Serial.println(batteryLevel);
  batteryLevelChar.writeValue(batteryLevel);  //update battery level
                                       //characteristic
  oldBatteryLevel = batteryLevel;      //save level for comparison
  }
}

//************************************************************
```

2.4 Application: Calibrating the Opta ADC

Scientific instruments potentially need calibration to a known benchmark. In this section we discuss calibration techniques that may be employed for a specific scenario. The techniques may be adapted for similar situations.

In this exercise, a 10 MOhm ten turn potentiometer was used to vary the input voltage to an Opta ADC input. One end of the potentiometer was connected to a 10 VDC supply and the other to ground. The potentiometer wiper contact was connected to the ADC input. The potentiometer was then adjusted for different input voltages ranging from 0 to 10 VDC at approximately 1 VDC intervals. Voltage readings from both an external calibrated digital voltmeter (DVM) was compared to the readings obtained from the Opta. Results are provided in Fig. 2.29. Note the percent difference between the readings.

To calibrate the Opta ADC, both the DVM and the Opta ADC voltage readings were plotted. Assuming that both devices provide a linear output (as evidenced by the plots), the output of the Opta ADC is first scaled by constant K. The scalar multiplier K maps the output range of the Opta ADC to the range of the DVM. Once the range has been mapped, the readings may now need to be shifted by a constant bias B.

Opta ADC value	Input Voltage DVM	Uncalibrated Opta ADC Voltage	Uncalibrated % Difference	Calibrated Opta Voltage	Calibrated % Difference
0	0	0	0	0	0
380	1	0.93	7.53	1.02	-2.00
756	1.99	1.85	7.57	2.03	-1.96
1139	2.99	2.78	7.55	3.05	-1.97
1585	4.15	3.87	7.24	4.25	-2.27
1837	4.83	4.49	7.57	4.93	-1.96
2358	6.21	5.73	8.38	6.29	-1.22
2740	7.21	6.72	7.29	7.37	-2.21
3193	8.39	7.8	7.56	8.56	-1.97
3433	9.09	8.38	8.47	9.19	-1.14
3665	9.82	8.95	9.72	9.82	0.00
K:	1.10				

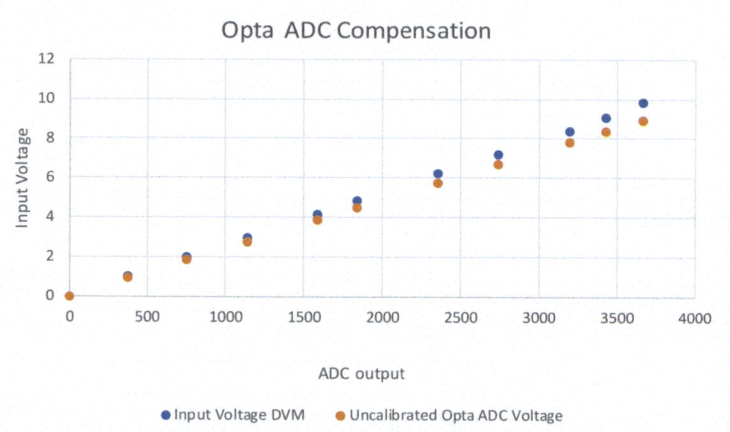

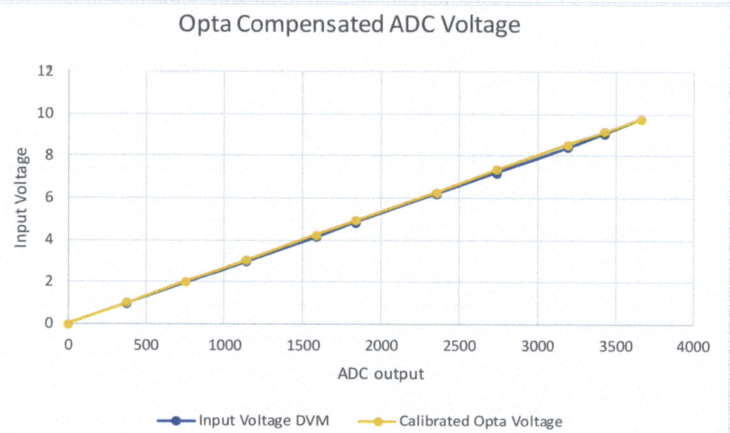

Fig. 2.29 Arduino opta ADC calibration

In general, the scaling and bias process may be described by two equations:

$$V_{DVMmax} = (V_{Optamax} \times K) + B$$

$$V_{DVMmin} = (V_{Optamin} \times K) + B$$

We thus have two equations and two unknowns to solve for K and B. These factors may be implemented within the Opta ADC sketch. For this calibration exercise K was equal to 1.10 V/V and B was 0. The resulting Opta calibrated voltage is shown with the corresponding reduction in percent difference.

Additional calibration techniques may be employed to further reduce the percent error between the calibrated and noncalibrated readings. For example, the noncalibrated values may be plotted versus the calibrated values using Excel. Employing Excel tools, a trendline may be established between the two sets of values. A line fitting equation (e.g. linear, polynomial, or power series) relating the two sets of values can be obtained. In general, the more complex the line fitting equation (e.g. power series), the lower the percent difference between the two sets of values.

2.5 Summary

We began the chapter with a brief review of the Arduino Opta series of micro PLCs. We explored features common to all three variants and employed them to explore fundamental input/output control concepts. We then examined and applied features specific to a given Arduino Opta variant. Throughout the chapter we provided illustrative examples. We concluded with a brief introduction to DIN rail-based implementation technology.

2.6 Problems

1. What is the difference between a microcontroller and a programmable logic controller?
2. Describe the Opta host processor features.
3. Construct a feature table for the Opta variants.
4. What are the characteristics of the Opta digital inputs and outputs.
5. What is a DIN rail? How is it used in industrial control applications?
6. What is switch bouncing? How is it corrected in PLC applications?
7. Create UML activity diagrams for all chapter sketches.
8. Describe different techniques for ADC calibration.
9. What is meant by the internet cloud?
10. What are the differences between a network configuration for a home versus a large business?
11. What is an ISP?

12. Draw the schematic of a home internet configuration using the Cisco System icons.
13. Draw the schematic of a large business internet configuration using the Cisco System icons.
14. In your own words write a brief summary of historical network development.
15. What is the difference between ARPANET, DARPANET, and NSFNET.
16. Summarize the differences between the ISO/OSI and the TCP/IP layered protocol models.
17. Describe the difference in how IPv4 and IPv6 addresses are allocated.
18. What is CIDR addressing? How does it extend the IPv4 addressing space?
19. What is the mission of the following agencies: ICANN, IETF
20. What is a MAC address? How is it different than an IP address?
21. What is the relationship between a DNS and an URL address?
22. Describe different sources of cybersecurity threats.
23. Describe measures to counter cybersecurity threats.
24. Provide a working definition of IoT and IIoT.
25. What is the difference between IT and OT? How are the concepts related.

References

Arduino Opta Collective Data Sheet–Product Reference Manual SKU: AFX00001–AFX00002–AFX00003, June 29, 2023, www.arduino.cc.

Bluetooth Low Energy, Wi–Fi, and Ethernet on Opta, www.opta.findernet.com.

M. Fowler with K. Scott "UML Distilled– A Brief Guide to the Standard Object Modeling Language," 2nd edition. Boston:Addison–Wesley, 2000. hero

Getting Started with Interrupts on Opta, www.opta.findernet.com.

Getting Started with Opta, www.opta.findernet.com.

Getting Started with RS–485 on Opta, www.opta.findernet.com.

Hanes D., G. Salgueiro, P. Grossetete, R. Barton, J. Henry (2017) *IoT Fundamentals–Networking Technologies, Protocols, and Use Cases for the Internet of Things*, Cisco Press.

Horowitz P, Hill W (2015) The Art of Electronics, third edition, Cambridge University Press.

Internet Corporation for Assigned Names and Numbers (ICANN), www.ICANN.org

Internet Engineering Task Force, www.IETF.org

Leiden C. and M. Wilensky (2009) TCP/IP for Dummies, 6th edition, John Wiley and Sons Publishing, Inc.

Levine R. and M. Levine Young (2015) *The Internet for Dummies*, John Wiley and Sons Publishing, Inc.

Lowe, D. (2018) Networking All–In–One for dummies, 7th edition, John Wiley and Sons Publishing, Inc.

Null L. and J. Lobur (2015) *Computer Organization and Architecture*, Jones and Bartlett Learning.
Opta User Manual, www.arduino.cc.
Shuler R., (2002) *How does the internet work?*, Rus Shuler @ Pomeroy IT Solutions.
Stenerson, J. (2004) *Fundamentals of Programmable Logic Controllers, Sensors, and Communications*, Pearson Prentice Hall.

Arduino PLC IDE and Ladder Logic

3

Objectives: After reading this chapter, the reader should be able to do the following:

- Describe and configure Arduino PLC programming software and related tools;
- List and describe five different methods of programming a PLC within the IEC IEC61131–3 standard;
- Summarize the fundamental concepts of Ladder Logic (LD) PLC programming;
- Describe library instructions within the Arduino PLC IDE; and
- Employ the Arduino PLC IDE to write, compile, and execute ladder logic programs for the Opta series of PLCs.

3.1 Overview

In this chapter we explore the Arduino PLC IDE. We begin with a review of the five PLC programming languages specified within the IEC IEC61131–3 standard and available within the Arduino PLC IDE. We then narrow our focus and concentrate on ladder logic programming. We explore, download, and configure the Arduino PLC IDE and related tools. We then explore the fundamentals of ladder logic PLC programming. The fundamentals are used to complete a series of examples.

S. F. Barrett, *Arduino VII*, Synthesis Lectures on Digital Circuits & Systems,
https://doi.org/10.1007/978-3-031-68609-2_3

3.2 Arduino Opta Programming Tools

In Chap. 2 we used the Arduino IDE to program various features and functions within the Opta PLC series. In this section we explore the Arduino PLC IDE and related software tools to program the Opta PLC series. The Arduino PLC IDE allows for programming Opta PLCs using the five different languages within the IEC IEC61131–3 standard. These include:

- Ladder Diagram (LD)
- Sequential (SFC)
- Function Block Diagram (FBD)
- Structured Text (ST)
- Instruction List (IL).

All of these languages are powerful and important; however, we concentrate on LD programming. In the next several sections we provide a brief introduction to this powerful, graphical programming technique.

3.3 Getting Started–Arduino PLC IDE

Complete the following steps to download and configure software tools, load configuration software to the Opta PLC, establish communication between the host laptop/PC, and activate the PLC IDE license[1]:

- **Software download and installation.** The Arduino PLC IDE and accompanying tools are available for download from arduino.cc. There are two different software packages required:

  - Arduino_PLC_IDE_Tools
  - Setup_Arduino_PLC_IDE

  Once downloaded, install the Arduino_PLC_IDE_Tools **first** followed by the Setup_Arduino_PLC_IDE.
- **Arduino PLC IDE**. With setup complete, open the Arduino PLC IDE by clicking on its icon. You should be greeted by the screen shown in Fig. 3.1a.
- Insert a project name, verify the target device is "Opta 1.0" and click "OK." A project window shown in Fig. 3.1b will appear.
- **Windows Device Manager**. With an Opta PLC connected to the support laptop/PC via the USB C cable, go to the Windows Device Manager to determine the two USB COM

[1] This section is adapted from "Arduino PLC IDE Setup & Device License Activation," arduino.cc.

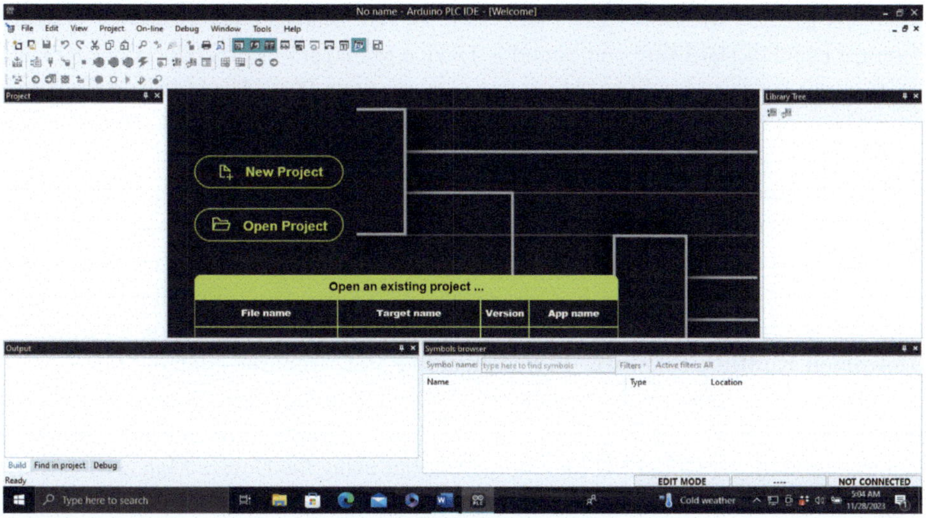

(a) Arduino PLC IDE welcome screen.

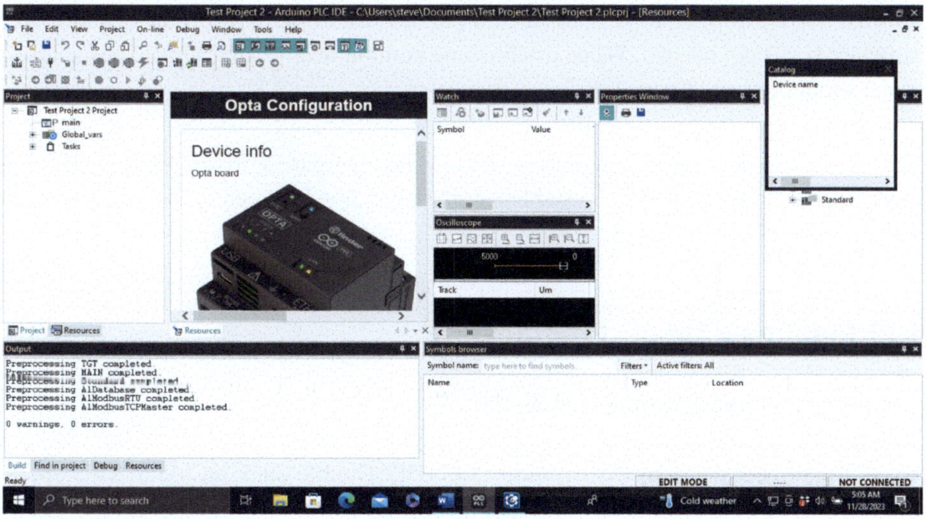

(b) New project screen.

Fig. 3.1 PLC IDE welcome screen. (Figure adapted and used with permission of Arduino team (CC BY–NC–SA) (www.arduino.cc).)

channels used by the Opta PLC. Access the Device Manager via Windows System – Control Panel – Hardware and Sound – Device Manager – Ports (COM & LPT). Two USB Serial Devices should be shown in use by the Opta PLC. The higher COM number is used by the MODBUS link and the lower number is the standard link. Note these two COM numbers for upcoming use.

- **Establish communication**. To connect the support laptop/PC to the Opta PLC select On–line – Set up communication. This opens up a new panel called "Device Link Manager Config." Choose "MODBUS" and "Properties." Verify MODBUS is using the higher COM number noted earlier. Close the panel with "OK." Go to On–line – connect to establish a communication link between the host laptop/PC and the Arduino Opta.
- **Download Runtime software**. To download the PLC Runtime configuration software to the Opta PLC, return to the Arduino PLC IDE. Go to the "Opta Configuration" panel and scroll to the bottom left corner. In the "Other" subpanel select the lower COM number used by the Opta PLC (determined earlier using the Device Manager) and click "Download." The Runtime sketch will compile and load to the PLC. This may take a little time. Progress may be monitored via the "Output" panel.
- **License Activation**. Opta is designated a pre–licensed PLC IDE product. Therefore, a software key is not needed. When establishing communication the "Opta Configuration" panel should reflect "Status OK."

This completes the steps required to download and configure software tools, load configuration software to the Opta PLC, establish communication between to the host laptop/PC, and activate the PLC IDE license.

3.4 Running a Simple Program

To compile and download a simple program:

- Go to the Project panel in the Arduino PLC IDE. Double click on Main. This opens a panel showing a simple counter program.
- Compile the program by going to Project – Compile.
- Download the compiled program to the Opta using On–Line – Download Code.
- The count (cnt) variable may be monitored using the Watch window. Go to View – Tool Windows – Watch. Configure the Watch window by double clicking on the Insert new item icon. A new panel will appear to insert the label name (cnt) and location (main).

This completes the software installation and configuration and running a first program. Note within the PLC IDE under Help is an extensive (250+ pages) "Arduino PLC User's Manual" for reference. In the next section we explore basic ladder logic concepts.

3.5 Structure of Arduino PLC IDE Program

Figure 3.2a demonstrates the flow of an Arduino PLC IDE program. The program executes the "Initialize" portion of the program once and then executes programs within each of the task categories (Fast, Slow, Background) at the time intervals shown.

Figure 3.2b shows program hierarchy within the Arduino PLC IDE. A project contains sections for Main, Global variables, and Tasks. The Task area is subdivided into categories previously described (Initialize, Fast, Slow, Background). Within each Task area are individual programs. A program consists of a sequence of networks (ladder rungs).

The network/ladder rung consists of a horizontal power link connecting the left power rail to the right power rail. Between the two power rails are the activities processed by the network. These include contacts (inputs) on the left and coils (outputs) on the right. The power link may also include additional activities from the Arduino IDE PLC Project Library as shown in Fig. 3.3.

A program is built up as a series of network/rungs to accomplish program steps. When the program is compiled and executed, the network/rungs are sequentially executed. Each network may be viewed as an IF–THEN statement. IF the contacts and the logic on the power link are TRUE–THEN the coil output is true. Let's take a closer look at ladder program components.

3.5.1 Contacts, Coils, Branches, and Blocks

Provided in Fig. 3.4 is a partial illustration of ladder logic components. The contacts are typically a normally open (NO) or normally closed (NC) momentary contact switch. As we see in upcoming examples, the contacts may also be control signals from prior ladder rungs.

Coils serve as the output for ladder networks (rungs). As seen in Fig. 3.4, control signals, coils maybe motors, lights, horns, etc.

Inputs maybe configured in series/parallel combinations. A series connection of input contacts imply a logical AND statement. That is both contacts must be at logical TRUE to assert the output coil. A parallel connection of input contacts imply a logical OR statement. In the parallel configuration either contact may be logical TRUE to assert the coil. The series and parallel contact configurations may be mixed to implement complex logic configurations.

The coil outputs may also be configured in a parallel configuration. For a parallel output coil configuration both coils are asserted simultaneously (e.g. a motor and indicator lamp).

Shown in Fig. 3.4 are the extensive library functions defined within the Arduino PLC IDE. Note the extensive operations, blocks, and standard functions. We use these in upcoming examples. Let's take a closer look at counters and timers.

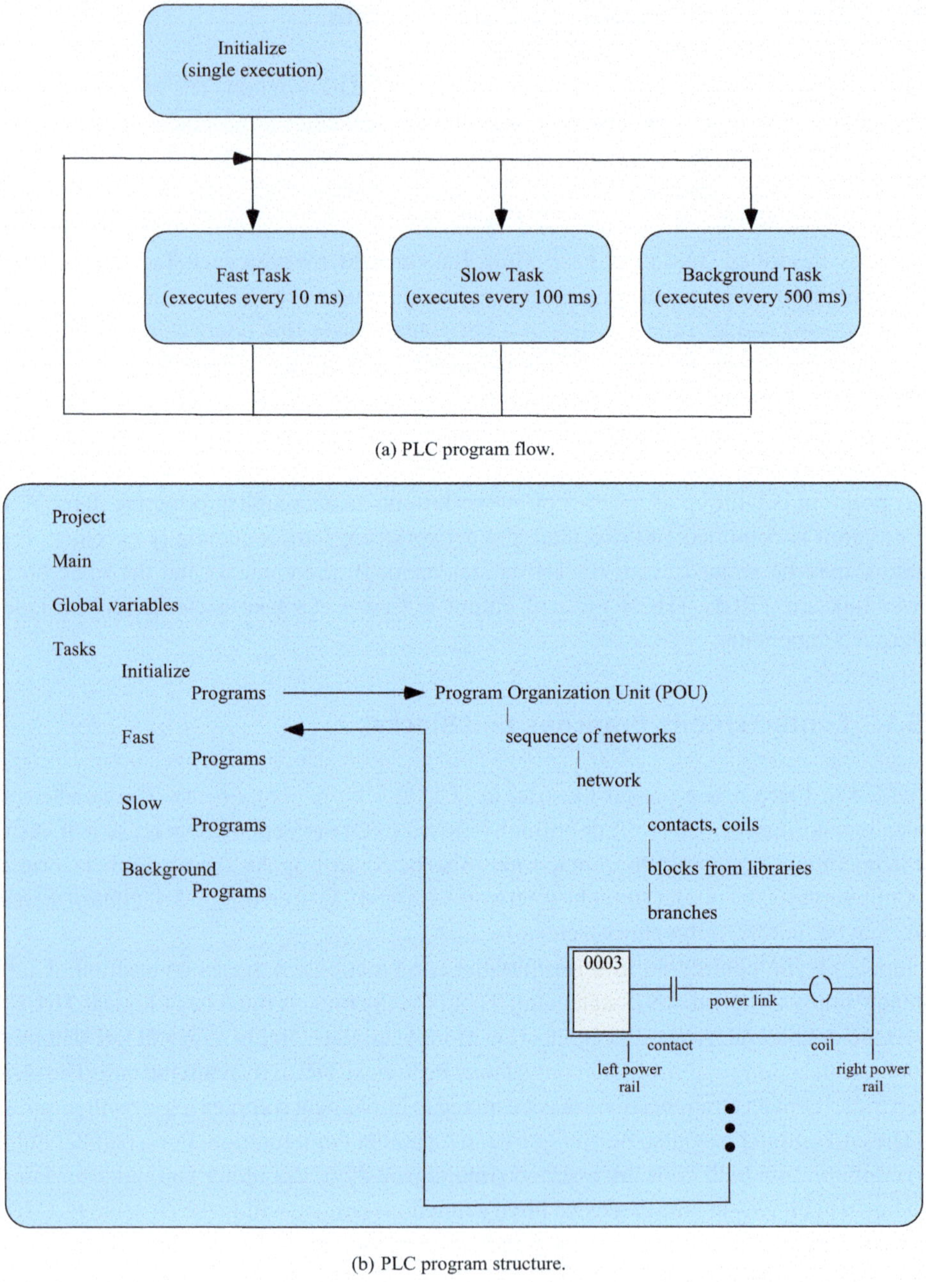

Fig. 3.2 Arduino PLC IDE program flow

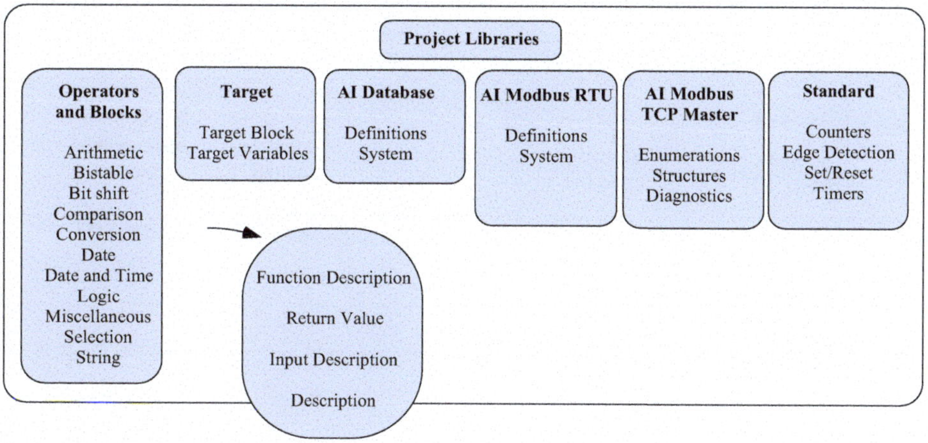

Fig. 3.3 Arduino PLC IDE project library

3.5.1.1 Counters

The Arduino PLC IDE features nine different predefined counters as shown in Fig. 3.5. There are counters that count down (CTD), count up (CTU), and up–down counters (CTUD). Each counter may either use signed 16 bit integers (INT), signed 32 bit integers (DINT), or unsigned 32 bit counters (DUINT). We explore the count–down timer (CTD) in some detail. Lessons learned may be applied to other counter types.[2]

The CTD timer is a signed, 16 bit counter. The CTD counts down by one (decrements) for each positive transition on the count–down (CD) input. The CTD has three inputs and two outputs:

- CD (Bool), input, count down on positive edge.
- LD (Bool), input, preset counter input, rising edge on LD sets Counter Value (CV) to Preset Value (PV).
- PV (Int), input, Preset Value, desired initial count of timer for count down.
- Q (Bool), output, count–down timer output. Set to logic one when counter value (CV) equals zero. Remains at zero until LD positive edge.
- CV (Int), output, current value of counter.

Reference Fig. 3.5 for a timing diagram illustrating counter operation.

---

[2] Detailed information for each counter type is accessible from within the Arduino PLC IDE Library.

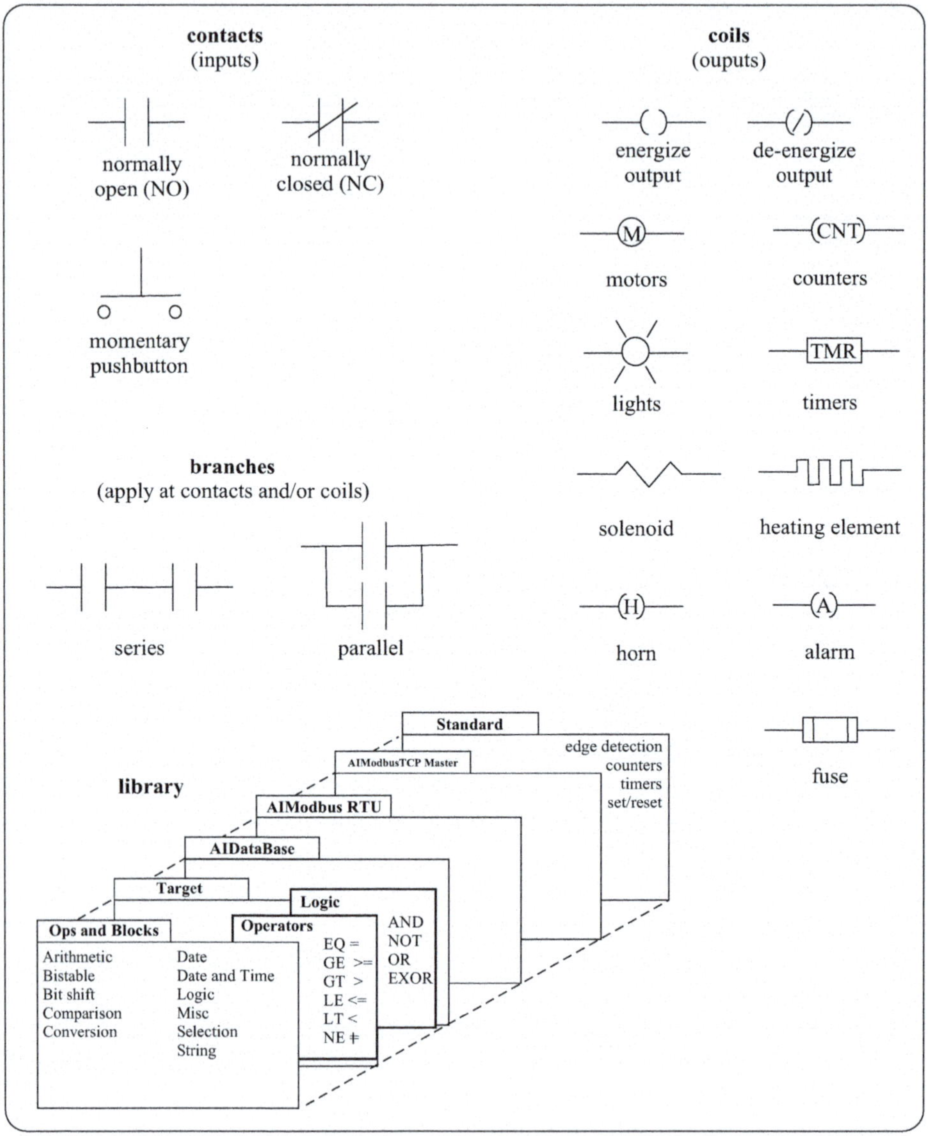

Fig. 3.4 Ladder logic network components

Counter Type	Counter Current Value Type INT	DINT	UDINT
down	CTD	CTD_DINT	CTD_UDINT
up	CTU	CTU_DINT	CTU_UDINT
up-down	CTUD	CTUD_DINT	CTUD_UDINT

Notes:
INT: signed, 16 bits
DINT : signed, 32 bits
DUINT: unsigned 32 bits

CTD - signed , 16 bit down counter
Counts down by one for each positive transition on the count down (CD) input.

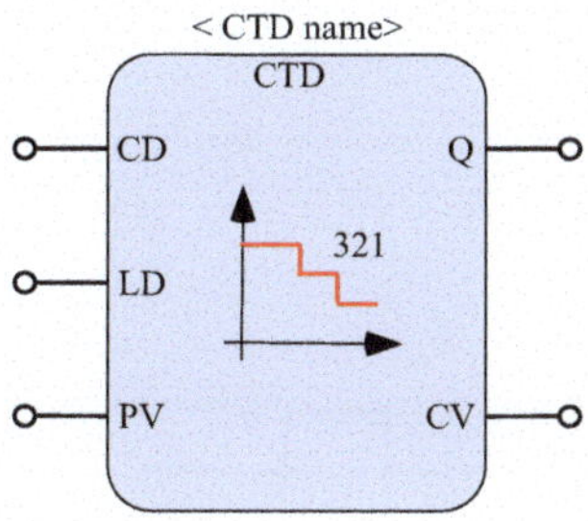

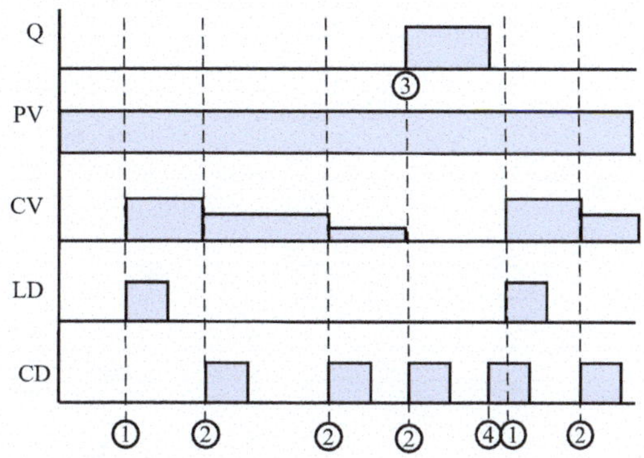

Q (BOOL): count-down output
PV (INT): preset value
CV (INT): counter current
LD (BOOL): preset counter input
CD (BOOL): count-down on positive edge

Notes:
1. Rising edge LD sets CV to value of PV.
2. Rising edge CD decrements CV.
3. When CV = 0, output Q set.
4. Q remains 0 until LD positive edge.

Fig. 3.5 Arduino PLC IDE counters

3.5.1.2 Timers

The Arduino PLC IDE features two different predefined timers: Off–delay Timer (TOF) and On–delay Timer (TON) as shown in Fig. 3.6. The TOF timer delays the deactivation of output Q by the preset time (PT) value in milliseconds. The TON timer provides a delay of PT milliseconds after IN becomes True before Q becomes True.[3]

3.5.2 LD Editor

Provided in Fig. 3.7 is a summary of launching a new ladder logic (LD) program within the Arduino PLC IDE and how to add and edit program rungs.

[3] Detailed information for each timer type is accessible from within the Arduino PLC IDE.

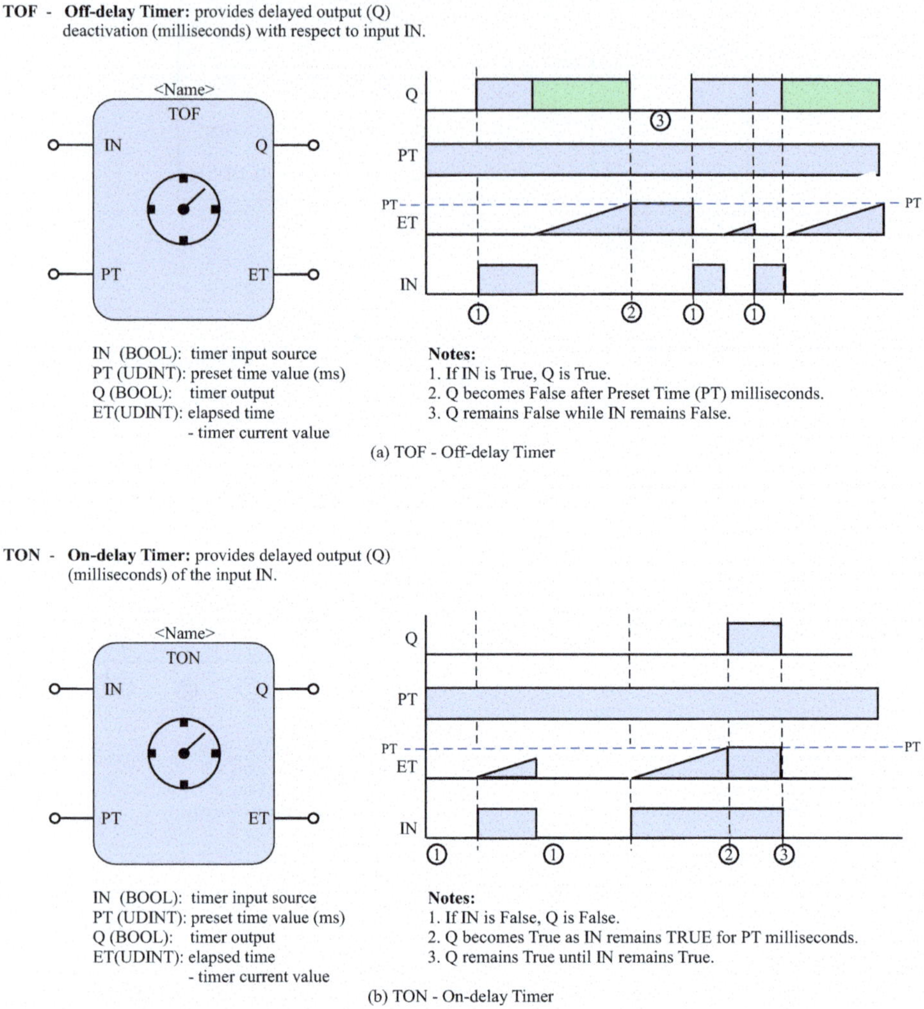

Fig. 3.6 Arduino PLC IDE timers

3.6 LD Program Examples

In this section we provide several examples to illustrate the use of ladder logic programming to provide basic control. Figure 3.8 provides the schematic used in the upcoming examples. In the Application section of the chapter we assemble the circuit into a DIN rail mounted Test Fixture.

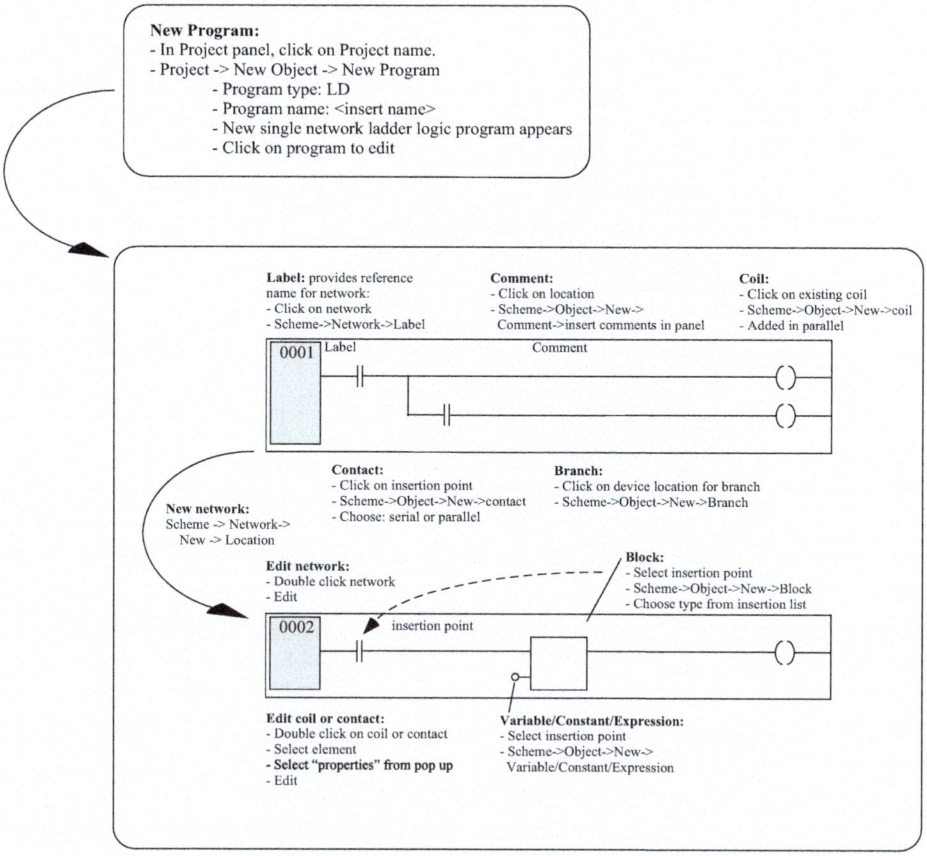

Fig. 3.7 Ladder logic network editing

Figure 3.9 provides the steps to launch an LD program. Follow the UML diagram to build the first ladder logic program. In the program, when the User pushbutton is pressed the first output relay is activated as shown in Fig. 3.10b.

Extend this basic example to implement and test the ladder logic circuits provided in Fig. 3.10a through d.

Figure 3.11 provides a basic counter configuration. In the example a count–down counter (CTD) is configured to start at the count of seven. The count is initially loaded by pressing PB1 connected to the preset counter input (LD). The CTD decrements for every push button press of PB2. When the counter reaches zero, output Q goes to logic high and illuminates the LED connected to Output_1.

Figure 3.12a provides a basic Off–delay Timer (TOF) configuration. In the example the TOF is configured with a two second (2000 ms) delay. When pushbutton PB3 is pressed the Q output goes logic high illuminating the LED at Output_2. When PB3 is released the

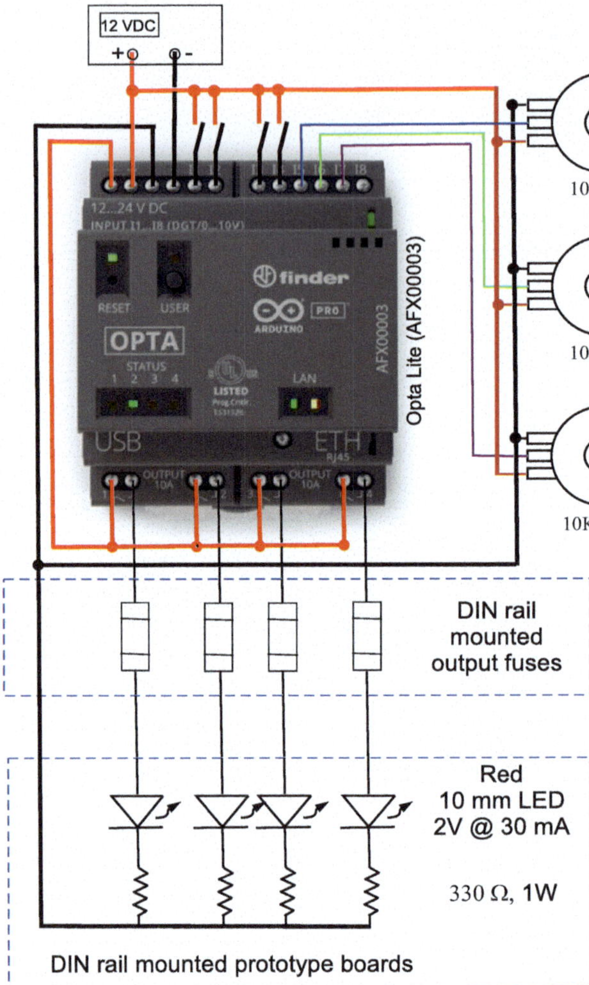

Fig. 3.8 Ladder logic test circuit. Images used courtesy of the Arduino Team (CC BY–NC–SA) (www.arduino.cc)

LED stays illuminated for another two seconds. In Fig. 3.12b note how the output Q from the count–down timer CTD1 serves as the enable signal for the TOF timer in place of PB3.

Figure 3.13 provides analog–to–digital (ADC) conversion example. A greater than (>) comparison operator is used to compare the analog voltage provided by a potentiometer connected to Input 5 (Pot 2) to a threshold value of 30,000. Input 5 is configured for 16–bit analog conversion so values range from 0 to 65,535 ($0\ to\ 2^{(b-1)}$) depending on the position of the potentiometer knob. When the Input 5 value exceeds 30,000 to LED connected to Output_2 illuminates. The Pot2 value may be observed using the Watch Window. The

Fig. 3.9 Ladder logic program

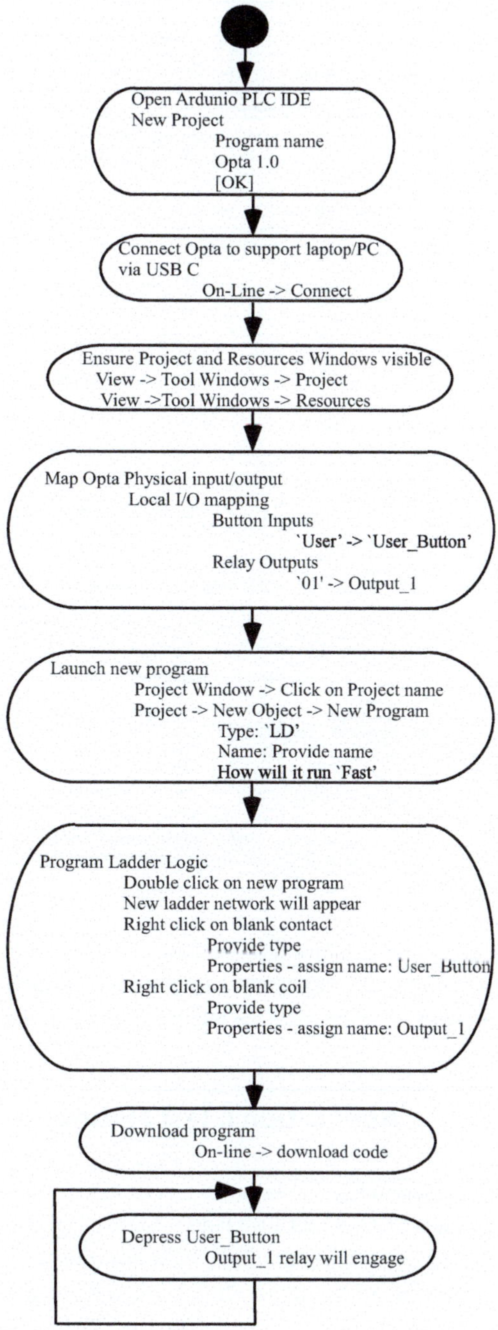

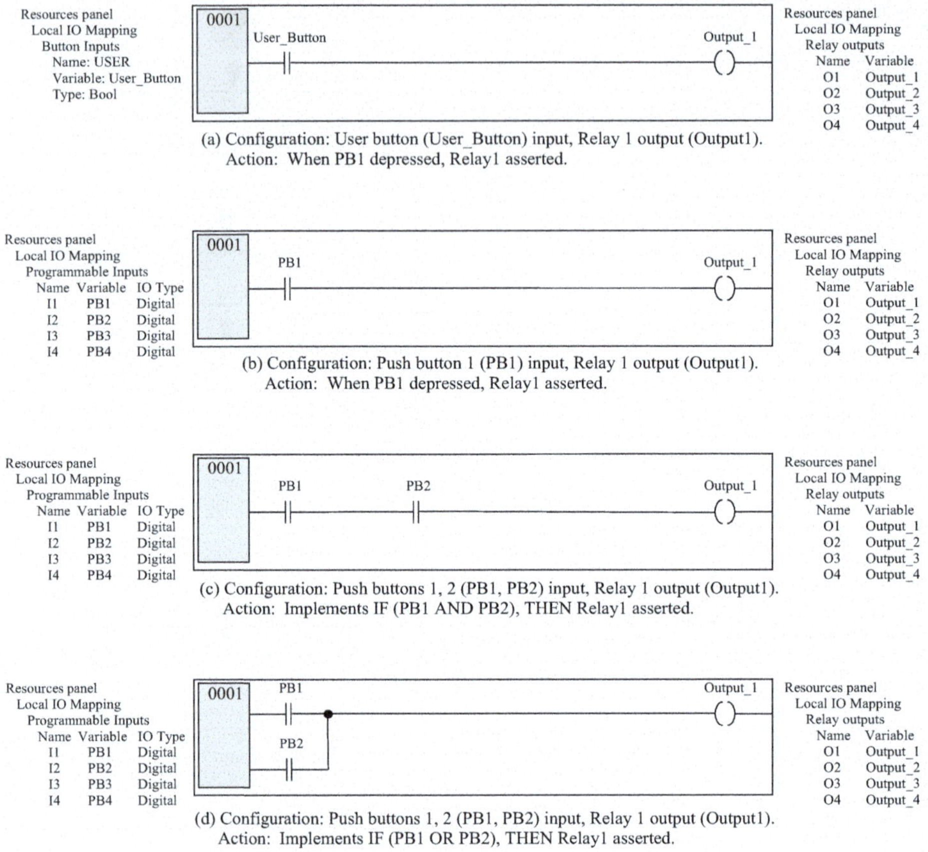

Fig. 3.10 Basic ladder logic examples

potentiometer represents any number of analog sensors (e.g. light, temperature, etc.). We discuss a wide variety of analog sensors in the next chapter.

3.7 Application I: Test Fixture

Provided in Fig. 3.14a is the test fixture to easily inject input signals and track outputs from an Opta PLC executing a ladder logic program. The test fixture consists of a series of three 10 Kohm potentiometers, four light emitting diodes (LEDs), and four pushbuttons configured as shown in Fig. 3.14b. The circuit is housed within a plastic chassis box mounted to a DIN rail via DIN rail mounting adaptors. The completed fixture is shown in 3.15.

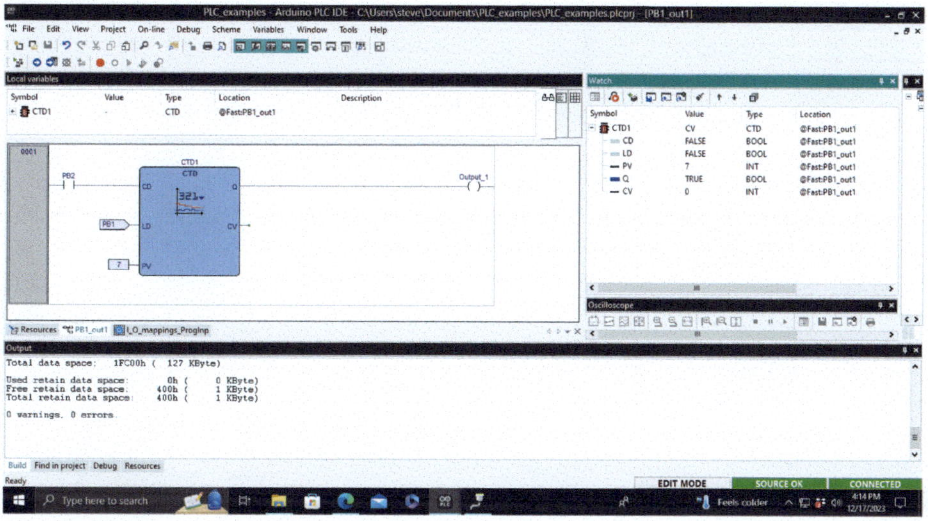

Fig. 3.11 Basic ladder logic counter example

3.8 Application II: Greenhouse Temperature Sensing System

Figure 3.16 provides a ladder logic implementation of a system to detect when a given physical variable (e.g. temperature, water level, etc.) is too high, too low, or within a desired range. We use the Test Fixture to test the ladder logic program.

Here is a summary of ladder logic operation:

- Resource definitions for Programmable Inputs and Relay Outputs are shown at the top left and right of the figure. Also, local variables of type Bool are declared for EN_Ladder, LE_OUT, and GE_OUT.
- Rung/network 0001 contains two contacts PB1 (NO) and PB2 (NC). The PB1 input serves as the Start PB for the process while PB2 serves as the STOP PB. When PB1 is pressed, the coil EN_Ladder is asserted and seals the PB1 switch ON via the parallel branch containing EN_Ladder. The process remains active until PB2 STOP is pushed.
- Rung/Network 0002 contains the greater than (>) comparison operator. It is used to compare the analog voltage provided by a potentiometer connected to Input 5 (Pot 1) to a threshold value of 60,000. Input 5 is configured for 16–bit analog conversion so values range from 0 to 65,535 ($0\ to\ 2^{(b-1)}$). When the Input 5 value exceeds 60,000 the Red LED illuminates. The Pot2 value may be observed using the Watch Window. It is important to note the potentiometer represents any number of analog sensors.

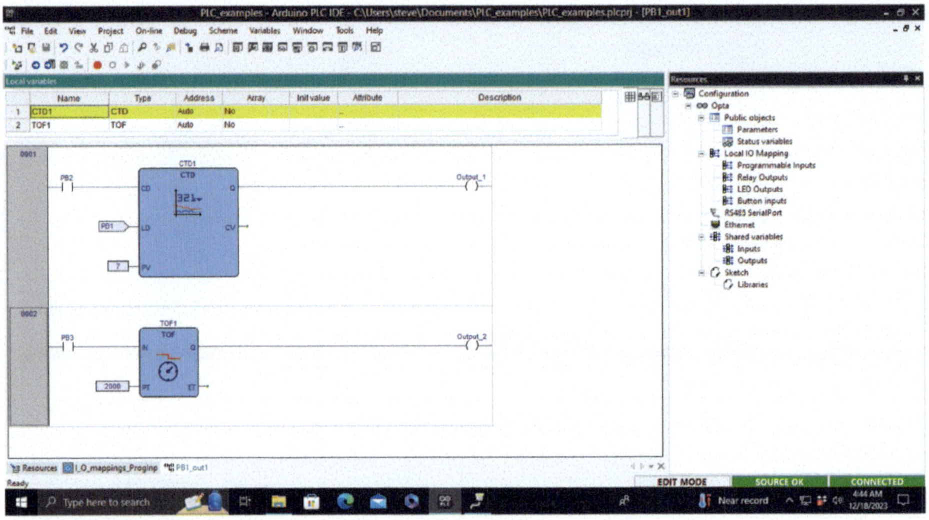

(a) TOF example.

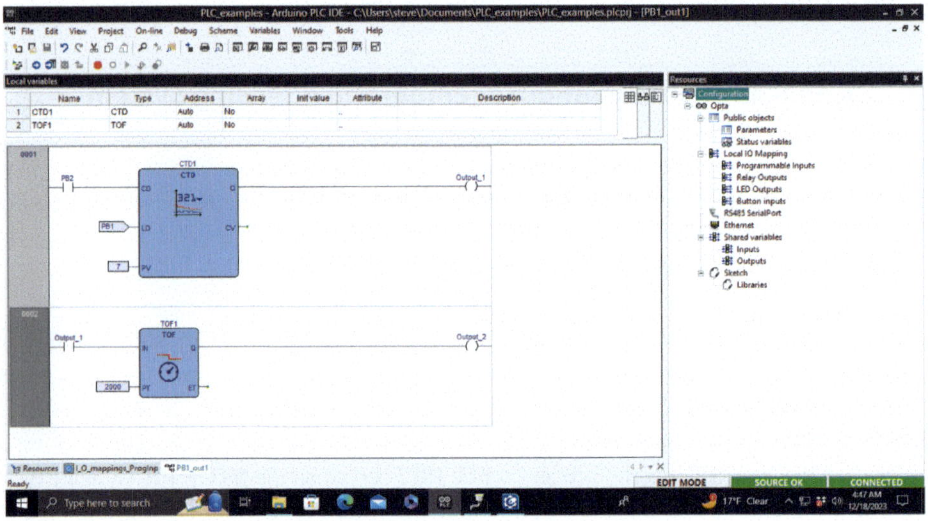

(b) Count down counter output as input for TOF.

Fig. 3.12 Basic ladder logic timer examples

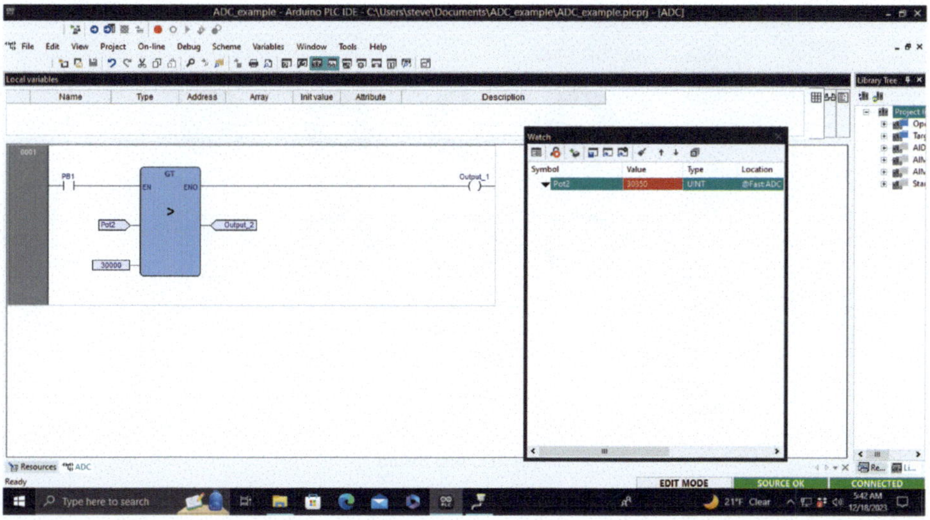

Fig. 3.13 Basic ladder logic ADC example

- Rung/Network 0003, 0004, and 0006 work together to illuminate a Green LED if the potentiometer connected to Input 5 (Pot 1) provides an output between 60,000 and 30,000. Note how the LE_OUT and the GE_OUT signals are ANDed together to form the Green_LED output.
- Rung/Network 0005 illuminate a Blue LED if the potentiometer connected to Input 5 (Pot 1) provides an output less than 30,000.

3.9 Summary

In this chapter we explored the Arduino PLC IDE. We began with a review of the five PLC programming languages specified within the IEC IEC61131–3 standard and available within the Arduino PLC IDE. We then narrowed our focus and concentrated on ladder logic programming. We explored, downloaded, and configured the Arduino PLC IDE and related tools. We then explored the fundamentals of ladder logic PLC programming. The fundamentals were used to complete a series of examples.

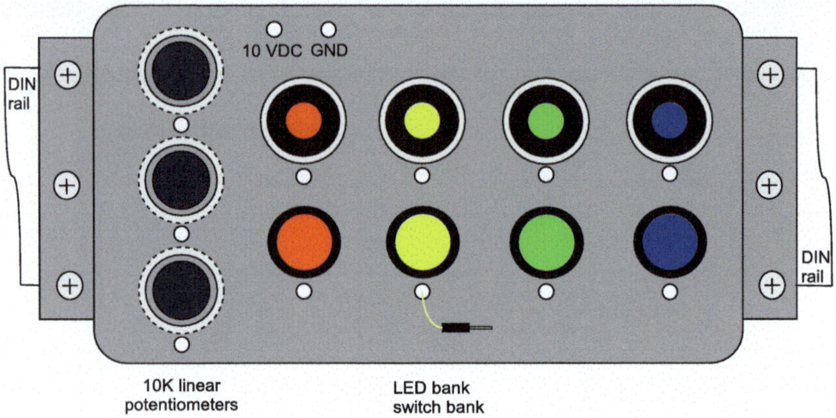

(a) PLC test fixture.

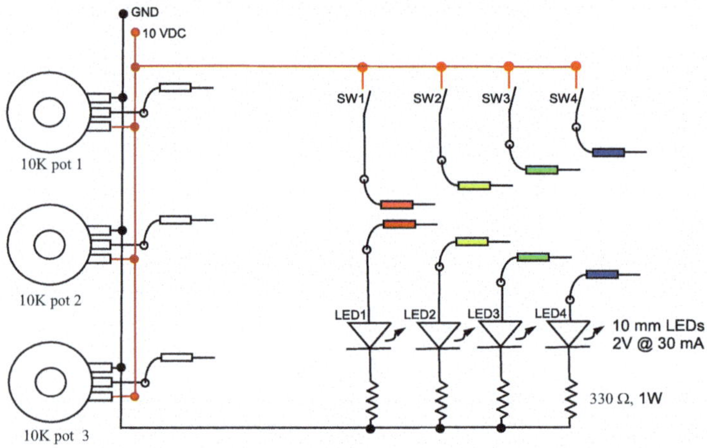

(b) PLC test fixture schematic.

Fig. 3.14 Ladder logic test fixture

3.10 Problems

1. The Arduino PLC IDE allows for programming Opta PLCs using the five different languages within the IEC IEC61131–3 standard. Prepare a table listing each type of programming language along with its pros and cons.[4]
2. Describe the difference between ladder logic contacts, coils, branches, and blocks. Provide an example of each.

[4] Accessible via the Arduino PLC IDE help tab.

Fig. 3.15 Ladder logic test fixture on DIN rail

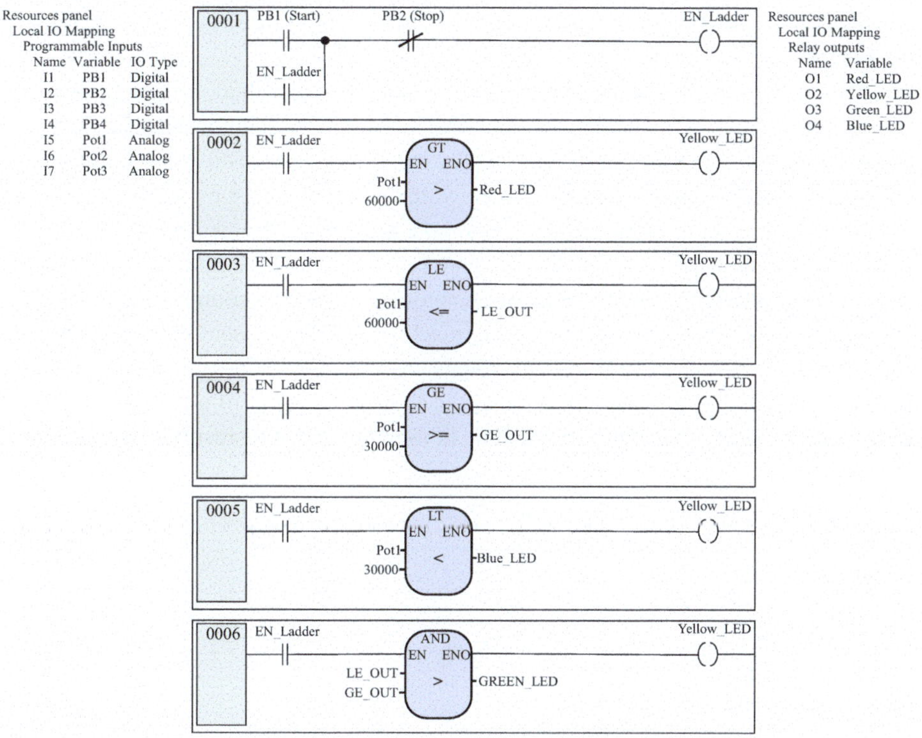

Fig. 3.16 Ladder logic sensing system

3. In Chap. 1 we provided an example to sequentially blink the four LEDs on the front of Opta PLC. Provide a ladder logic program to accomplish the sequential blinking with 500 ms delays.
4. Modify the ladder logic program above so the LEDs sequentially cycle left to right and then back right to left.
5. There are many predefined operations, blocks, and standard functions defined with the PLC IDE Library. Provide a summary table of these library elements.
6. Describe the difference between the TOF and the TON timers. Write a ladder logic program to demonstrate the difference between the two timers.
7. Design a ladder logic program providing a signal when 12 cans on a conveyor has passed a sensor. The signal is used to start a new box. The boxes are packed 144 per palette. Provide a signal when 144 boxes have been filled with cans.
8. Develop a ladder logic circuit to indicate the fluid level in a vat. Provide output signals when the vat is nearing empty or nearing full. An analog sensor is used to indicate vat fluid level.
9. A small corporation is owned by four individuals. The distribution of the corporation shares is as follows: Mr. Quine owns four shares, Mrs. Karnaugh owns three shares, Mr. Boole owns two shares, and Ms. McCluskey owns one share. You are to build a voting machine that will be used when the owners vote on corporate issues. Each of the owners has a switch that is closed to record a 'yes' vote and opened to record a no' vote. When one of the owners casts a yes' vote, that vote is weighted by the number of the owner's shares. For example, when Mrs. Karnaugh votes yes', her vote counts three times that of Ms. McCluskey. The output of the voting machine will be a visual indication–maybe a LED that is lit? When the result of any vote is a yes' by the majority of the shareholders. If the issue does not get a majority vote, the indicator LED will not light. If the issue results in a tie vote, a separate LED will be lit to indicate this tie result. Develop a ladder logic program to implement the voting machine.[5]

References

Arduino Opta Collective Datasheet, Product Reference Manual, SKU: AFX00001– AFX00002–AFX00003, arduino.cc, June 2023.

Arduino PLC User's Manual, arduino.cc.

P. Marquinez, *Arduino PLC IDE Setup & Device License Activation,* arduino.cc.

Programming Introduction with Arduino PLC IDE–Create Programs with all the IEC–61131–3 languages on the Arduino PLC IDE, Finder S.p.A., opta.findernet.com.

J. Bagur and J.C. Linares, *Opta User Manual,* arduino.cc

[5] Hint: Employ Boolean logic simplification techniques (e.g. Karnaugh maps) to determine equations for a successful vote and tie. Implement the equations with PLC IDE logical operators. Scenario courtesy of Alex Dwellis, Colonel, USAF retired.

Input Sensors, Output Actuators, and Interfacing 4

Objectives: After reading this chapter, the reader should be able to:

- Apply the voltage and current input/output parameters toward properly interfacing input sensors and output actuators to an Arduino Opta PLC;
- Interface a wide variety of input and output devices to an Arduino Opta PLC;
- Describe operational amplifier (op amp) function, ideal characteristics, and common configurations;
- Apply op amp concepts in the design of a transducer interface; and
- Apply chapter concepts interface multiple input and output devices to the Opta PLC.

4.1 Overview

The Opta PLC may be used in industrial, Internet of Things (IoT), and hybrid applications. We find sensors and actuators at a variety of common voltages (e.g. 5 VDC). In this chapter we explore how to connect these input sensors and output actuators to an Opta PLC. We begin with a review of the Opta input and output characteristics. We then explore a wide variety of digital and analog input sensors and output actuators. We employ an operational amplifier–based transducer interface design process to interface input sensors to the Opta PLC.[1]

[1] The information on embedded system design first appeared in "Microcontroller Fundamentals for Engineers and Scientists," Morgan and Claypool Publishers, 2006. It has been adapted with permission. Although first developed for embedded systems design, concepts provided here apply to Opta–based PLC system design.

S. F. Barrett, *Arduino VII*, Synthesis Lectures on Digital Circuits & Systems,
https://doi.org/10.1007/978-3-031-68609-2_4

4.2 Opta Input and Output Operating Parameters

In Chap. 2 we provided a summary of Opta PLC inputs and relay outputs. This information is provided here for convenience.

4.2.1 Programmable Digital/Analog Inputs

The Arduino Opta series is equipped with eight inputs designated I1–I8. Each input can be configured as a digital input or an analog input. In Chaps. 2 and 3 we described how to configure the inputs using the Arduino IDE and the Arduino PLC IDE. We discuss each configuration in turn (Opta).

4.2.1.1 Digital Input

The Arduino Opta input pins are designated I1–I8 on the PLC case. A digital input value may range from 0 to 24 VDC. Values less than 4.46 VDC are considered logic low while those greater than 6.6 VDC are considered logic high. The inputs are equipped with over voltage protection.

4.2.1.2 Analog Inputs

When configured as an analog input, the input signal ranging from 0 to 10 VDC is converted to a corresponding digital value. The corresponding digital value may be configured for 12–16 bits of resolution.

The maximum allowable voltage to the Arduino Opta PLC is 10 VDC. However, the internal host PLC processor performing the ADC has a maximum allowable voltage of 3 VDC. Therefore, the voltage for ADC conversion provided to the Opta PLC input is scaled internally by a factor of 0.30. When a measurement is taken, the result must be rescaled by this value.

4.2.2 Relay Outputs

The Arduino Opta is equipped with four normally open (NO) mechanical relay contacts. The relay contacts are rated at 250 VAC, 10A. Since they are AC contacts, current can flow in either direction through the relay. The relay outputs are not fused. The relay outputs have the following characteristics (Opta):

- Response time from logic 0 to 1: 6 ms
- Response time from logic 1 to 0: 4 ms
- Bounce time NO: 3 ms
- Bounce time NC: 6 ms

4.3 Input Sensors

A PLC is typically used in applications where data is collected by input sensors, the data is then processed by the host algorithm, and then a control decision and accompanying signals are provided by the PLC to output peripheral actuators. The input sensors may be digital or analog in nature.

4.3.1 Digital Input Sensors

Digital sensors provide a single digital input signal or a series of digital logic pulses with sensor data encoded. The sensor data may be encoded in any of the parameters associated with a digital signal such as logic level, duty cycle, frequency, period, pulse length, or pulse rate. In the next several sections we describe how to interface a wide variety of digital input devices including switches and sensors to the Arduino Opta PLC. We begin with switches.

4.3.2 Switches

Switches come in a variety of types. A system designer chooses the appropriate switch for a specific application. Switch varieties commonly used in control applications are illustrated in Fig. 4.1. Here is a brief summary of the different types:

- **Slide switch**: A slide switch has two different positions: on and off. The switch is manually moved to one position or the other. For control applications, slide switches are available in a panel mounted package. Also, small slide switches are available that fit in the profile of a common integrated circuit size dual inline package (DIP). A bank of four or eight DIP switches in a single package is commonly available. Slide switches may be used to select specific parameters at system startup.

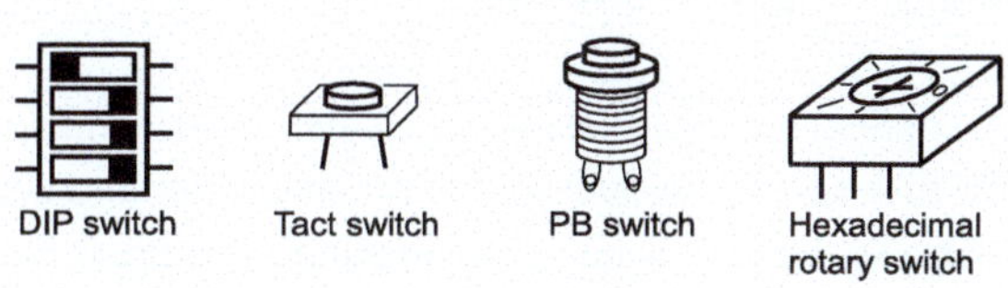

Fig. 4.1 Switch types

- **Momentary contact pushbutton switch**: A momentary contact pushbutton switch comes in two varieties: normally opened (NO) and normally closed (NC). A normally open switch, as its name implies, does not normally provide an electrical connection between its contacts. When the pushbutton portion of the switch is depressed, the connection between the two switch contacts is made. The connection is held while the switch is depressed. When the push button is released, the connection is opened. The opposite is true for a normally closed switch. For control applications, pushbutton switches are available in a small tact type switch configuration or panel/chassis mounted configurations. Posh button style switches are commonly used in ladder logic coding to start and stop a control process.
- **Push on/push off switches**: This type of switch is also available in a normally open or normally closed configuration. For the normally open configuration, the switch is depressed to make connection between the two switch contacts. The pushbutton must be depressed again to release the electrical connection.
- **Hexadecimal rotary switches**: Small profile rotary switches are available for controller applications. These switches commonly have sixteen rotary switch positions. As the switch is rotated to each position, a unique four–bit binary code is provided at the switch contacts. These switches are used to select specific parameters at system startup.

4.3.2.1 Switch Debouncing

Mechanical switches do not make a clean transition from one position (on) to another (off). When a switch is moved from one position to another, it makes and breaks contact multiple times. This activity may go on for tens of milliseconds. A processor such as the Opta PLC is relatively fast as compared to the action of the switch. Therefore, the processor may recognize each switch bounce as a separate and erroneous transition.

To correct the switch bounce phenomena additional external hardware components may be used or software techniques may be employed. Software switch debouncing is accomplished by inserting a 30–50 ms lockout delay in the function responding to input changes. The delay prevents the processor from responding to the multiple switch transitions related to bouncing.

The programming and interface between a switch and the Opta PLC is described in Chaps. 2 and 3.

4.3.3 Optical Encoder

An optical encoder consists of a small plastic transparent disk with opaque lines etched into the disk surface. A stationary optical emitter and detector pair is placed on either side of the disk. As the disk rotates, the opaque lines break the continuity between the optical source

Fig. 4.2 Optical encoder

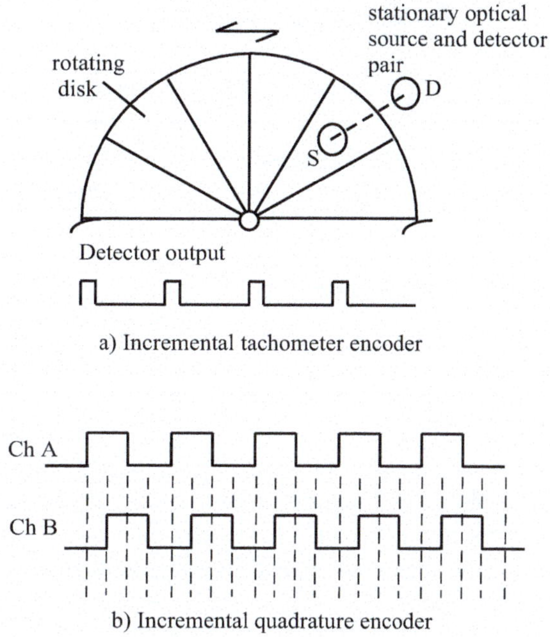

and detector. The signal from the optical detector is monitored to determine the rate of disk rotation as shown in Fig. 4.2.

Optical encoders are available in a variety of types depending on the information desired. There are two major types of optical encoders: incremental encoders and absolute encoders.

An absolute encoder is used when it is required to retain position information when power is lost. The absolute encoder is equipped with multiple data tracks to determine the precise location of the encoder disk (Sick Stegmann).

An incremental encoder is used in applications where a velocity or a velocity and direction information is required. The incremental encoder types may be further subdivided into tachometers and quadrature encoders. An incremental tachometer encoder consists of a single track of etched opaque lines as shown in Fig. 4.2a. It is used when the velocity of a rotating device is required. To calculate velocity, the number of detector pulses are counted in a fixed amount of time. Since the number of pulses per encoder revolution is known, velocity may be calculated.

The quadrature encoder contains two tracks shifted in relationship to one another by 90°. This allows the calculation of both velocity and direction. To determine direction, one would monitor the phase relationship between Channel A and Channel B as shown in Fig. 4.2b.

In the Application section we explore motor speed control using an optical tachometer.

4.4 Analog Input Sensors

Analog input sensors provide a DC voltage that is proportional to the physical parameter being measured. The analog voltage is converted to a corresponding binary representation. With analog sensors, signal preprocessing may be required to convert the sensor output to an analog DC voltage suitable for measurement by the Opta PLC.

4.4.1 Flex Sensor

An analog flex sensor is shown in Fig. 4.3a. The flex sensor provides a change in resistance for a change in sensor flexure. At 0° flex, the sensor provides 10 k Ω of resistance. For 90° flex, the sensor provides 30–40 k Ω of resistance. Since the Opta PLC cannot measure resistance directly, the change in flex sensor resistance must be converted to a change in a DC voltage. This is accomplished using the voltage divider network shown in Fig. 4.3c. For increased flex, the DC voltage will increase. The voltage and hence flex can be measured using the analog–to–digital converter subsystem.

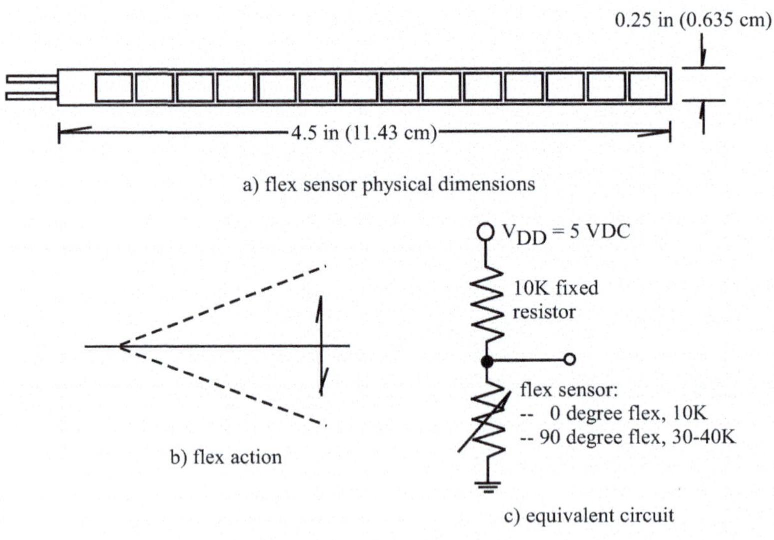

Fig. 4.3 Flex sensor

4.4.2 Ultrasound Sensor

The ultrasonic sensor pictured in Fig. 4.4 is based on the concept of ultrasound or sound waves that are at a frequency above the human range of hearing (20 Hz to 20 kHz). The ultrasonic sensor pictured in Fig. 4.4c emits a sound wave at 42 kHz. The sound wave reflects from a solid surface and returns back to the sensor. The amount of time for the sound wave to transit to the surface and back to the sensor may be used to determine the range from the sensor to the wall.

Pictured in Fig. 4.4c and d is an ultrasonic sensor manufactured by Maxbotix, the LV–Max Sonar—EZ series, MB1010–000. The sensor provides an output that is linearly related to range in three different formats: (a) a serial RS–232 compatible output at 9600 bits per second, (b) a pulse width modulated (PWM) output at a 147 us/inch duty cycle, and (c)

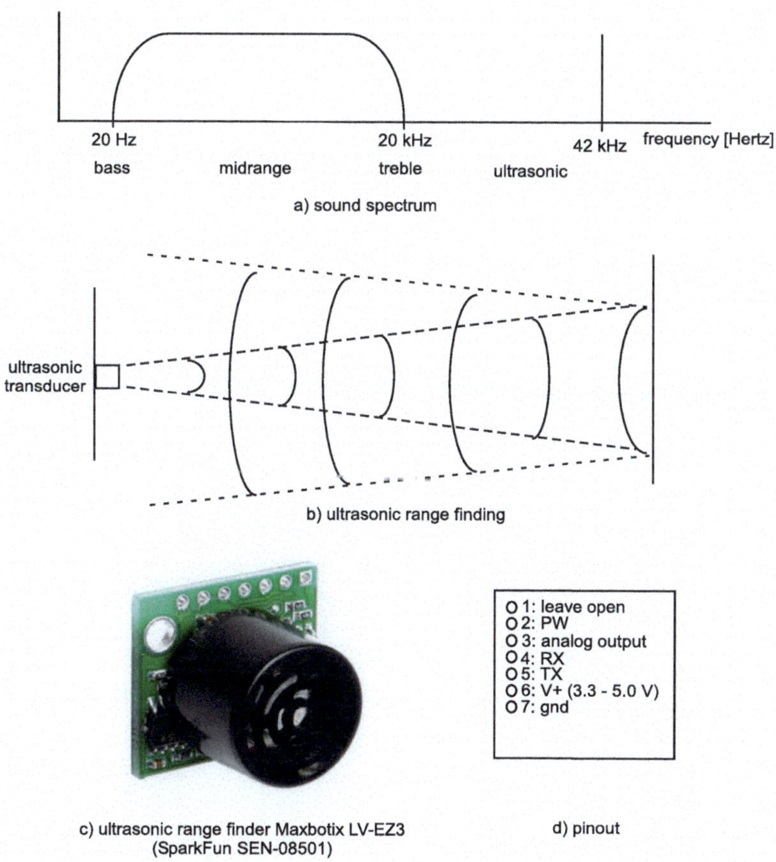

Fig. 4.4 Ultrasonic sensor. Sensor image used courtesy of SparkFun electronics (CC BY–NC–SA) (www.sparkfun.com)

an analog output at a resolution of approximately 10 mV/inch when Vcc is 5.0 VDC. The sensor is powered from a 2.5 to 5.5 VDC source (www.sparkfun.com).

In this example we use the sensor's analog output to determine range.

```
//*************************************************************
//ultrasonic - demonstration of Maxbotix LV-MaxSonar-EZ
//             MB1010
//Connections:
//MB1010:
//  +5 VDC
//  Ground
//  Analog Output to Opta I1
//  Commom ground between MB1010 and Opta
//Source: \url{www.maxbotix.com}
//Adapted for use with Opta PLC inputs - S. Barrett, 12.27.23
//*************************************************************

int       sensorValueA0;              //from ultrasound sensor
float     inches, voltageA0;

void setup()
{
Serial.begin(9600);
analogReadResolution(12);             //set 12 to 16 bits
}

void loop()
{
//Read the input on analog input I1 corresponding to A0:
sensorValueA0 = analogRead(A0);
float voltageA0 = (float)(sensorValueA0) * (10.0/4095.0);

//Print out value from I1
Serial.print("I1 value: ");
Serial.print(sensorValueA0);
Serial.print(" corresponding to ");

//Print voltage as float with 2 decimal digits
Serial.print(voltageA0, 4);
Serial.println(" Volts");
//inches = (float) (sensorValueA0) * 5.0/25.4;
//inches = (sensorValueA0)/0.009766;                //to inches
inches = voltageA0 * 102.4;

//Print range
Serial.print("Range:");
Serial.print(" ");
Serial.print(inches);
```

```
Serial.println(" inches");
delay(2000);
}

//*****************************************************************
```

4.4.3 Temperature Sensors

There are several sensor types that may be used to measure temperature including thermocouples, thermistors, and integrated circuit (IC) based sensors. Due to their linearity and ease of use, we concentrate on the IC based sensors. Provided in Fig. 4.5 is a summary of common integrated circuit (IC) based temperature sensors. The sensors may be used over a wide temperature range and provide an output voltage as a linear function of temperature. The sensors are available in an IC type or small transistor style packaging.

Example: LM34 Temperature Sensor. In this example we use the LM34 Precision Fahrenheit Temperature Sensors manufactured by Texas Instruments to measure temperature. The LM34 provides 10 mV of output per degree Fahrenheit. The output pin of the LM34 is provided to input pin I1 (A0) on the Arduino Opta PLC for temperature readings. Provided below is an Arduino IDE based sketch to measure the output from the LM34, convert the LM34 output to temperature, and display the result on the Serial Monitor.[2]

Temperature Sensor	Temperature Range	Accuracy over Range	Scale Factor	Supply Voltage
LM34	-50° to 300° F	+/- 1 to 1/2°F	10.0 mV/°F	+5 to +20 V*
LM35	-55° to 150° C	+/- 1 to 1/2°C	10.0 mV/°C	+4 to +20 V*
TMP35	10° to 125° C	+/- 2°C	10.0 mV/°C	+2.7 to +5.5V
TMP36	-40° to 125° C	+/- 2°C	10.0 mV/°C	+2.7 to +5.5V
TMP37	5° to 100° C	+/- 2°C	20.0 mV/°C	+2.7 to +5.5V

*Requires negative supply for temperature readings below 0

Fig. 4.5 Integrated circuit temperature sensors. (LM34, LM35 www.ti.com, TMP35, 36, 37 temperature sensors www.analog.com)

[2] As discussed in Chap. 2, calibration of the ADC system with a known calibrated source may be required.

```
//*************************************************************
//Opta_LM34
//
//Measures temperature using LM34
// - LM34 provides 10 mV/degree F
//
//S. Barrett, 12.27.23
//This example code is in the public domain.

int       sensorValueA0;              //from ultrasound sensor
float     degrees, voltageA0;

void setup()
{
Serial.begin(9600);
analogReadResolution(12);            //set 12 to 16 bits
}

void loop()
{
//Read the input on analog input I1 corresponding to A0:
sensorValueA0 = analogRead(A0);
float voltageA0 = (float)(sensorValueA0) * (10.0/4095.0);

//Print out value from I1
Serial.print("I1 value: ");
Serial.print(sensorValueA0);
Serial.print(" corresponding to ");

//Print voltage as float with 2 decimal digits
Serial.print(voltageA0, 2);
Serial.println(" Volts");
degrees = voltageA0 * 100.0;

//Print range
Serial.print("Temp:");
Serial.print(" ");
Serial.print(degrees);
Serial.println(" degrees (F)");
delay(2000);
}

//*************************************************************
```

Example: LM34 Temperature Sensor with ladder logic. In this example an LM34 is used to measure ambient temperature. When the temperature exceeds 90° Fahrenheit, a fan turns on. The fan remains on until the ambient temperature drops below 70° Fahrenheit. The difference between the fan on setting and off setting is called hysteresis. The two values are different to ensure the fan is not constantly turning on and off near the set point. The circuit diagram is provided in Fig. 4.6 and the accompanying ladder logic program in Fig. 4.7.

The LM34 temperature sensor output is fed directly to Opta input I3. The fan and LEDs (Y, G, B) are connected to Opta relay outputs 1–4. The thresholds to turn the fan on and off are determined by:

- Equating the temperature reading to LM34 output (e.g. 90° provides 900 mV).
- Determining the output from a 12–bit ADC for this voltage.
- The ADC output at the specified voltage is the threshold.

$$threshold = (0.900/10.0) \times 4096) = 370$$

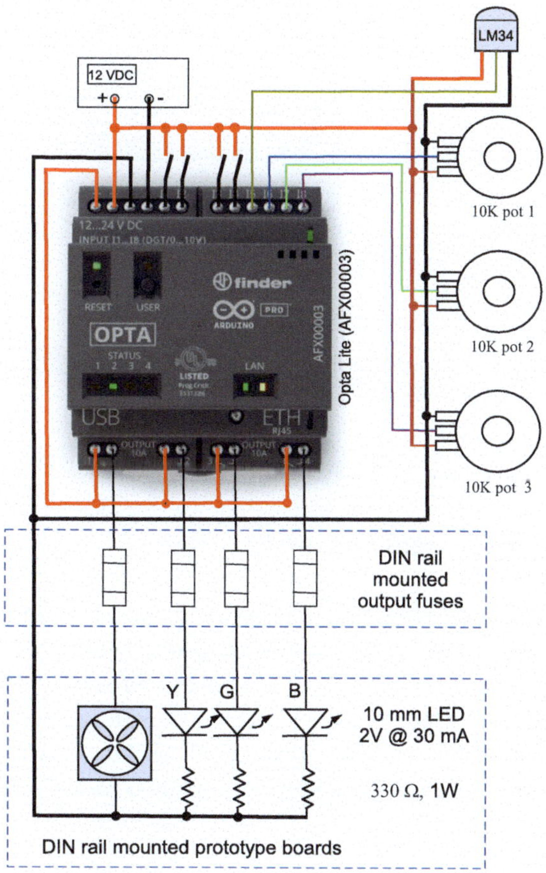

Fig. 4.6 LM34 fan controller. Images used courtesy of the Arduino Team (CC BY–NC–SA) (www.arduino.cc)

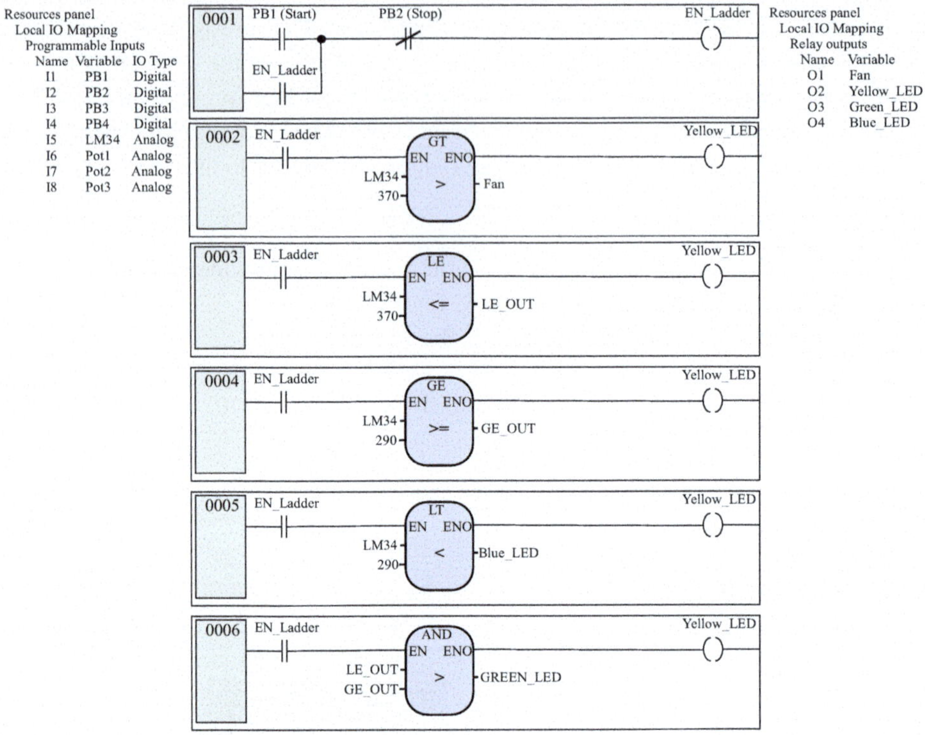

Notes:
1. ADC inputs set for 12 bits.
2. ADC resolution: 10 VDC/2^{12} = 2.44 mV/step
3. LM34 provides 10.0 mV/°F
4. LM34 provides output of 900 mV at 90°F. With a 12 bit ADC this corresponds to an ADC output of 370.
5. LM34 provides output of 700 mV at 70°F. With a 12 bit ADC this corresponds to an ADC output of 290.

Fig. 4.7 LM34 fan controller ladder logic

4.4.4 Light Sensor

There are many different types of sensors used to detect light including photoresistors, photovoltaic cells, photodiodes, and phototransistors. When choosing a light sensor it is important to match the sensor characteristics to the desired wavelength of light. Shown in Fig. 4.8 is a portion of the electromagnetic spectrum. The visible light spectrum ranges from approximately 380–760 nm. Each visible color has a defined range of wavelengths. The ultraviolet and infrared spectrums are adjacent to the visible spectrums.

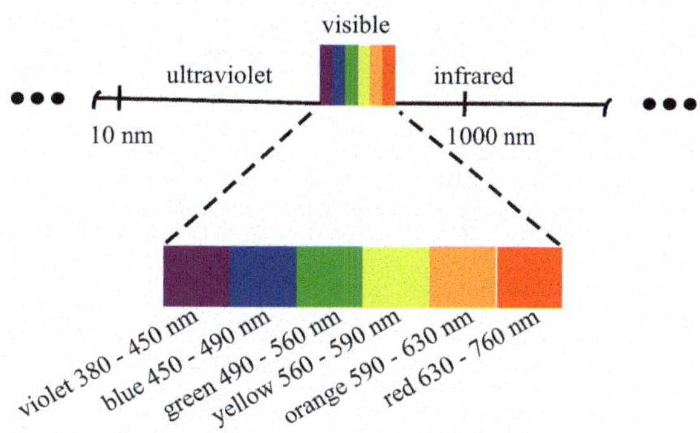

Fig. 4.8 Light spectrum (Miller)

4.4.4.1 Photoresistor

A photoresistor may be known by several different names including photoconductive cells, photo cells, and light dependent resistors. They are constructed from different types of semiconductor material each responsive to different light wavelengths in the infrared (IR), visible, or ultraviolet (UV) bands. When light of the appropriate wavelength impinges on the semiconductor material, current carriers are released and the resistance of the material is decreased. The photoresistor may be used in applications to detect day and night, light or dark environments, or to detect the presence of an object.

Example: PDV–P8001 Cadmium Sulfide Cell. In this example we use a PDV–P8001 Cadmium Sulfide (CdS) photoconductive photocell to determine when night has arrived. The cell is sensitive to light in the visible range (400–700 nm). The resistance of the cell ranges from 200 k Ω when in darkness to approximately 400 Ω when fully illuminated (API).

To interface to the Arduino Opta a 100 k Ω resistor is placed in series with the cell as shown in Fig. 4.9a to implement a voltage divider circuit. The node between the photocell and the resistor is connected to Arduino Opta ADC input I1 (A0). The threshold for determining day versus night is experimentally determined. The results are provided to the Serial Monitor for display. The example sketch was created using the Arduino IDE.

```
//*********************************************************************
//opta_photocell
// - Reads analog voltage at A0 connected to PDV-P8001 circuit
// - Photocell in series with 100k Ohm resistor
//S. Barrett, 01.05.24
//This example code is in the public domain.

int photo_reading;
int photo_threshold = 50;

void setup()
{
Serial.begin(9600);
analogReadResolution(12);          //set 12 to 16 bits
```

```
}

void loop()
{
//Read the input on analog input I1 corresponding to A0:
photo_reading = analogRead(A0);

//Print out value from I1
Serial.print("Photo sensor reading: ");
Serial.println(photo_reading);

if(photo_reading > photo_threshold)//night
  {
  Serial.println("Night");
  }
else
  {
  Serial.println("Day");
  }
delay(2000);
}

//*********************************************************************
```

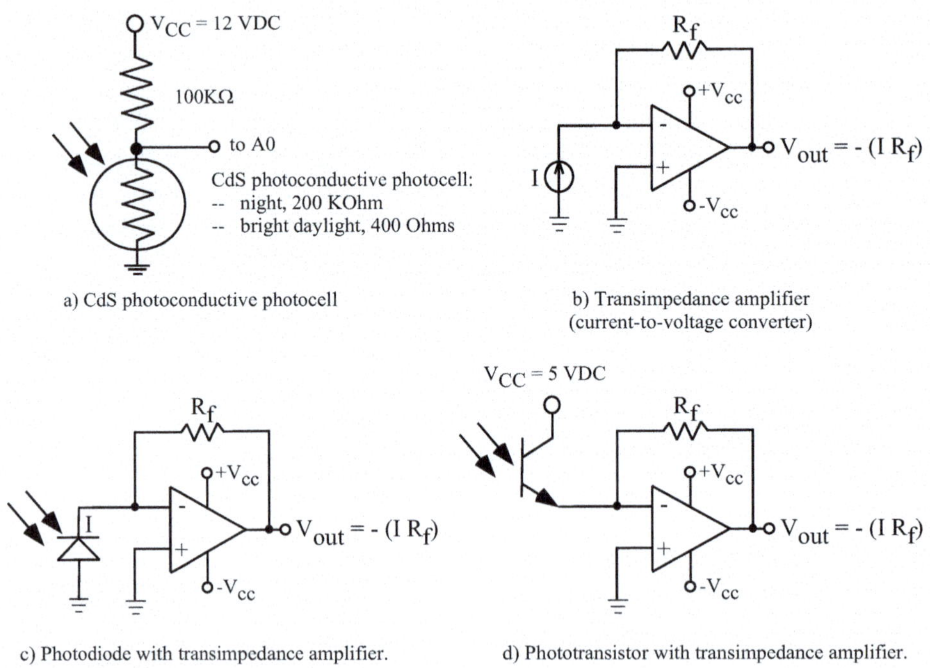

Fig. 4.9 Light sensors

4.4.4.2 Photodiode and Phototransistor

The photodiode and phototransistor are semiconductor devices that generate current in the presence of light. They are typically used with an op amp based transimpedance amplifier (Fig. 4.9b) to convert the current to a voltage.[3]

The photodiode (Fig. 4.9c) is used in the reverse bias or photoconductive mode. The generated current flowing through the feedback resistor (R_f) provides the output voltage (V_{out}). The phototransistor generates more current but responds slower in response to light changes.

Optical links are used to isolate one circuit from another, provide a communication link, or provide isolation in an industrial application. An optical link may be formed by using an LED coupled with a photodiode or phototransistor responsive to the LED wavelength via an optical fiber.

Example: Optical fiber link. Industrial Fiber Optics manufacturers a series of optical emitters (LEDs) at a wide variety of wavelengths (e.g. IF–E93 (green), IF–E97 (red), IF–E92B (blue)). They also manufacture a wide variety of optical detectors including photo diodes (e.g. IF–D91B), transistors (e.g. IF–D92B), and Darlingtons (e.g. IF–D93B) (Industrial Fiber Optics).

In this application we use an optical fiber to couple a light emitting diode to a photodiode to form an optical data link. We start with some background information on optical fibers. Optical fibers are used to link two devices via light rather than an electronic signal. In a typical application an electronic signal is converted to light, transmitted down the optical fiber, and converted back to an electronic signal.

As shown in Fig. 4.10a an optical fiber consists of several concentric layers of material including the core where light is transmitted, the cladding, the buffer, and the protective outer jacket. Light is transmitted through the fiber via the concept of total internal reflection. The core material is more optically dense than the cladding material. At shallow entry angles the light reflects from the core/cladding boundary and stays within the fiber core as shown in (b). This allows for the transmission of light via fiber for long distances with limited signal degradation.

To provide an interface between an electronic signal and the fiber, an optical emitter is used as shown in (c). The optical emitter contains a light emitting diode (LED) as the light source. At the far end of the optical fiber an optical detector is used to convert the light signal back to an electronic one.

As previously mentioned, it is important to note that optical emitters, detectors, and fibers are available in a variety of wavelengths. It is essential that the emitter, detector, and fiber are capable at operating at the same optical wavelengths.

[3] Operational amplifiers are discussed later in the chapter.

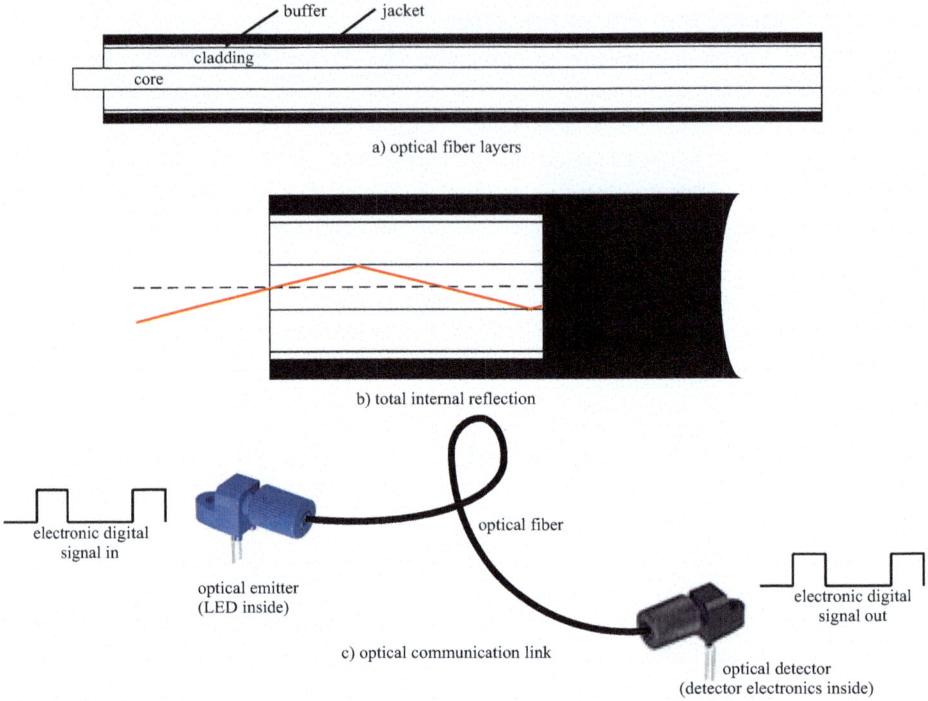

Fig. 4.10 Optical fibers

Example: Optical fiber link. In this example we use a transistor (PN2222) to switch the red LED (660 nm, IF-E97, $V_f = 1.7$, I_f 40 mA) on/off. The transistor is driven by an Opto PLC relay output as shown in Fig. 4.11. On the receiving end we use a photodiode (IF–91B) with a transimpedance amplifier to convert the light signal back to an electronic signal.

We use an Arduino IDE based sketch to generate a digital signal. The resulting input and output waveforms for the optical link are shown in Fig. 4.11. The 0 VDC reference point for each signal is shown as cursors on the left side of the image. Note the signal inversion from input to output. Later in the chapter we explore how to use operational amplifier building blocks to condition the signal to desired characteristics.

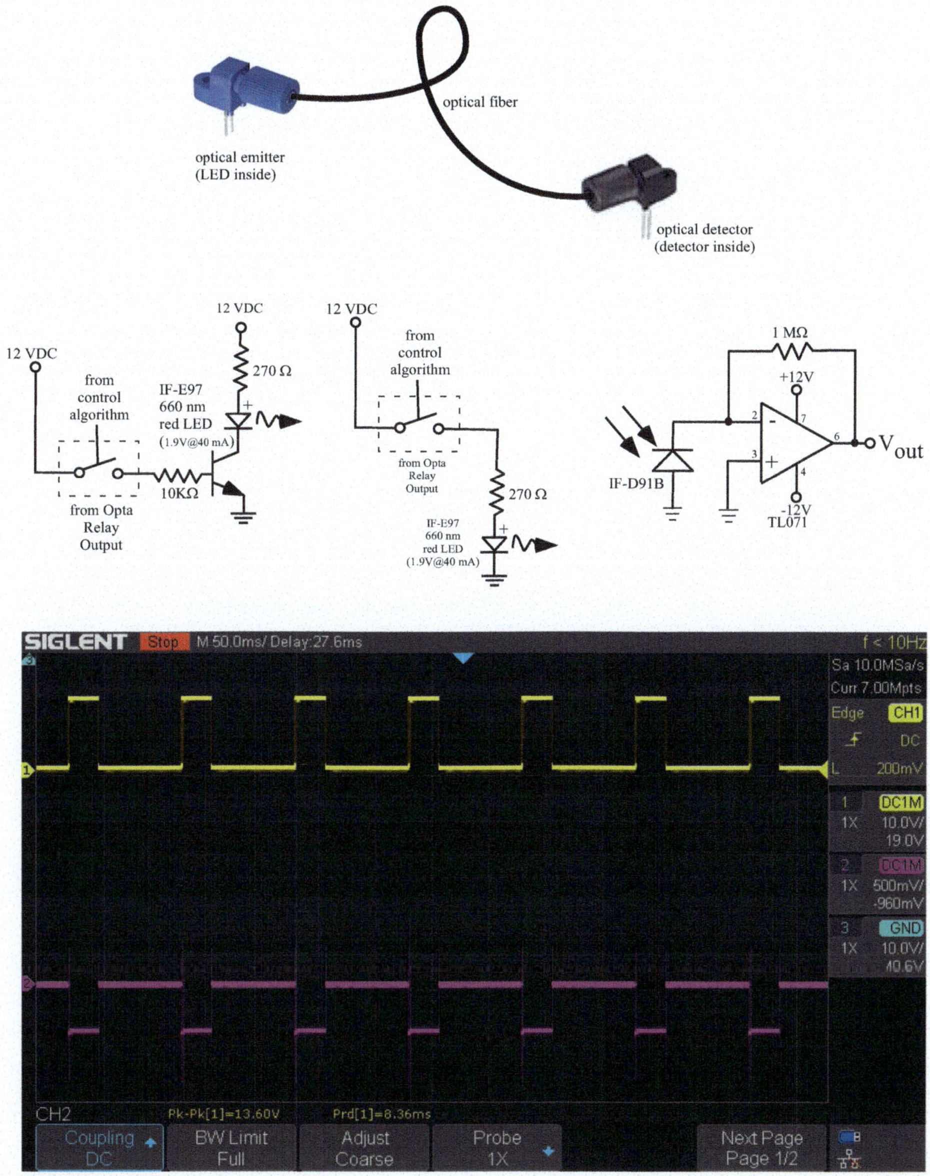

Fig. 4.11 Interface for optical fibers

```
//**************************************************************************
//Opta_opto_link
//
//S. Barrett, 01.15.24
//This example code is in the public domain.

int           opto_high_time, opto_low_time;

void setup()
{
Serial.begin(9600);                         //serial monitor BAUD rate
pinMode(D0, OUTPUT);                        //Initialize Relays outputs
pinMode(LED_D0, OUTPUT);                    //Initialize Opta LEDs
}

void loop()
{
digitalWrite(D0, HIGH);                     //Relay 1 on
digitalWrite(LED_D0, HIGH);                 //LED 1 on
delay(25);                                   //PWM: high baseline

digitalWrite(D0, LOW);                     //Relay 1 off
digitalWrite(LED_D0, LOW);                 //LED 1 off
delay(75);                                //PWM: low baseline
}

//*******************************************************************
```

4.4.5 Tilt Sensor

CTi Sensors manufactures a series of dual–axis inclinometers equipped with three–access accelerometers. The sensor outputs may either be digital or analog. We explore the TILT–15–S–90 that provides ±90° inclinometer sensing for both the X and Y axis. The sensor may be powered from 4.1 to 38 VDC. We use a 12 VDC supply to power the sensor. We also use the sensor's analog output. The sensor provides 25 mV per degree over the full ±90° range. The analog output voltage ranges from 0.25 V to 4.75 VDC (www.CTiSensors.com).

Example: Tilt sensor In this example we use a CTi TILT–15–X–90 sensor to measure the angle between a stationary and movable strut as shown in Fig. 4.12. We use the sensor's analog OutX output. The sensor provides 0.25V at minus 90° tilt and 4.75V at plus 90°. It has a sensitivity of 25 mV per degree at tilt measurements between the two extremes.

The OutX signal is fed to Arduino Opta input I1 (A0). The value from the ADC converter is mapped to an angular value. At 0.25 V the ADC output is approximately 51; whereas, at 4.75 V the ADC output is approximately 972. We use these values with the Arduino map function to convert the ADC readings to an angular value between minus 90 and plus 90°.

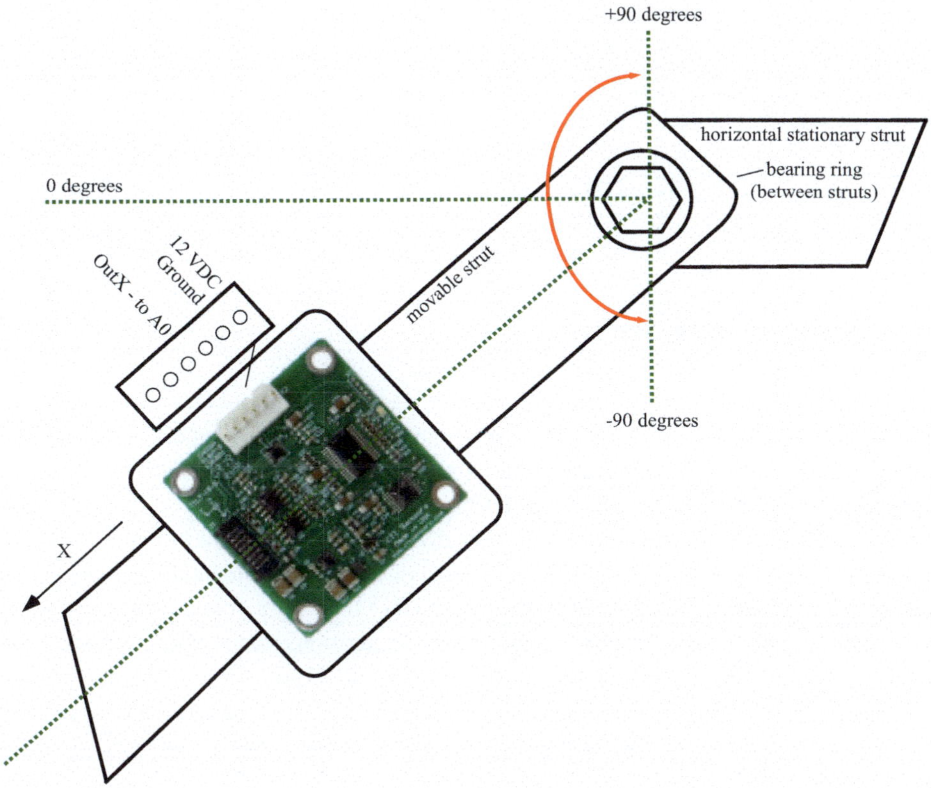

Fig. 4.12 CTi TILT–15–S–90 sensor (www.CTiSensors.com)

```
//**********************************************************************
//tilt_test:
//- CTi tilt sensor analog output OUTX is connected to Arduino
//  UNO R3 pin A0.
//- The sketch reads ADC value at A0, maps to degree, and displays
//  value on the Serial Monitor every second.
//This example code is in the public domain.
int analogPin = A0;                     //sensor OUTX
 int value=0;                           //value read from A0
 int tilt;                              //angular displacement

void setup()
{
Serial.begin(9600);                     //configure Serial Monitor
}

void loop()
{
value = analogRead(analogPin);          //read the input pin
Serial.print("Sensor reading: ");
```

```
Serial.println(value);              //sensor reading
tilt = map(value, 51, 972, -90, 90);
Serial.print("Tilt reading: ");
Serial.println(tilt);               //tilt value
Serial.println(" ");
delay(1000);
}
//*****************************************************************
```

4.4.6 Environmental Sensors

In this section we concentrate on the MQ series of gas and environmental sensors. The MQ series of sensors consists of a metal oxide semiconductor active element. The electrical resistance of the active element varies in the presence of a specific gas or gases. A sample of available MQ sensors is shown in Fig. 4.13 (www.mysensors.org).

An MQ sensor is typically used in a voltage divider circuit as shown in Fig. 4.14a. In the presence of a specific gas or gases, the resistance of the sensing element (R_S) varies. The value of R_S is in series with a fixed load resistor (R_L) forming a voltage divider network. The output voltage is an indication of the gas concentration. Many of the sensors in the MQ series require a heater voltage as shown in Fig. 4.14b. Figure 4.14c shows the physical configuration of the MQ series of sensors. SparkFun provides a breakout board to allow an MQ sensor to interface with a standard prototype board (www.mysensors.org, www.SparkFun.com).

To develop an interface circuit for an MQ sensor, complete the following steps:

- Choose a specific detectable gas of interest.
- Choose an appropriate sensor from Fig. 4.13.
- Determine key interface parameters as provided in Fig. 4.13.
- Implement the interface circuit as shown in Fig. 4.14b.
- Select an appropriate value of load resistor (R_L). A suggested value of load resistance is provided in the sensor's data sheet.[4] It is recommended to use a potentiometer that includes the value of suggested load resistance as the load resistor. This will allow the adjustment of circuit sensitivity to a specific level of gas concentration.
- Write an Arduino sketch to read the analog output voltage from the sensor interface circuit, set a threshold for detection, and illuminate an LED and sound a buzzer when the gas is detected.

Example. MQ sensor. In this example we develop an interface circuit and an Arduino Opta PLC IDE sketch as a smoke detector. We use the MQ–4 sensor as the active element. The

[4] Data sheets for many of the MQ sensors are provided at www.mysensors.org.

MQ-2	liquefied petroleum gas (LPG), propane, hydrogen, methane V_C < 24V, V_H = 5V, R_H = 31 ohms, P_H < 900 mW, preheat: 48 hours
MQ-3	alcohol, benzene, methane, hexane, LPG, carbon monoxide V_C = 5V, V_H = 5V, R_H = 33 ohms, P_H < 750 mW, preheat: 24 hours
MQ-4	high sensitivity to methane and natural gas, lower sensitivity to alcohol and smoke V_C = 5V, V_H = 5V, R_H = 33 ohms, P_H < 750 mW, preheat: 24 hours
MQ-6	liquefied petroleum gas (LPG) V_C < 24V, V_H = 5V, R_H = 26 ohms, P_H < 950 mW, preheat: 24 hours
MQ-7	carbon monoxide V_C = 5V, R_H = 33 ohms, P_H < 350 mW, preheat: 48 hours V_H : alternates between 5V for 60s and 1.4V for 90s
MQ-8	hydrogen V_C = 5V, V_H = 5V, R_H = 29 ohms, P_H < 900 mW, preheat: 48 hours
MQ-9	methane, propane, carbon monoxide V_C <10V, R_H = 31 ohms, P_H < 350 mW, preheat: 48 hours V_H : alternates between 5V for 60s and 1.5V for 90s
MQ-131	ozone V_C = 5V, V_H = 6V, R_H = 31 ohms, P_H < 1,100 mW, preheat: 24 hours
MQ-135	ammonia, nitrogen oxide, alochol, benzene, smoke, carbon dioxide V_C = 5V, V_H = 5V, R_H = 33 ohms, P_H < 800 mW, preheat: 24 hours
MQ-138	hexane, benzene, ammonia, alcohol, smoke, carbon monoxide V_C = 5V, V_H = 5V, R_H = 31 ohms, P_H < 800 mW, preheat: 24 hours
MQ-214	methane, LPG, butane, propane V_C = 6V

Fig. 4.13 MQ sensor series. Data sheets for many of the MQ sensors are provided at www.mysensors.org

manufacturer's data sheet recommends a load resistance of 20 k Ω. A 100 k Ω potentiometer, set for 20 k Ω, serves as the load resistor as shown in Fig. 4.15 or a fixed resistor may be used. In clear air Vout is below 1.0 V. In the presence of smoke Vout approaches 2.0 V.

The ladder logic sketch provided in Fig. 4.16 senses Vout and activates a Sonalert when smoke is detected at a set sensitivity level. The threshold value for activation is set for Vout at 1.0 VDC.

The MQ series of sensors may be calibrated to provide a reading in gas parts per million, or "PPM." The interested reader is referred to procedures in the appropriate data sheet (Egironi 4).

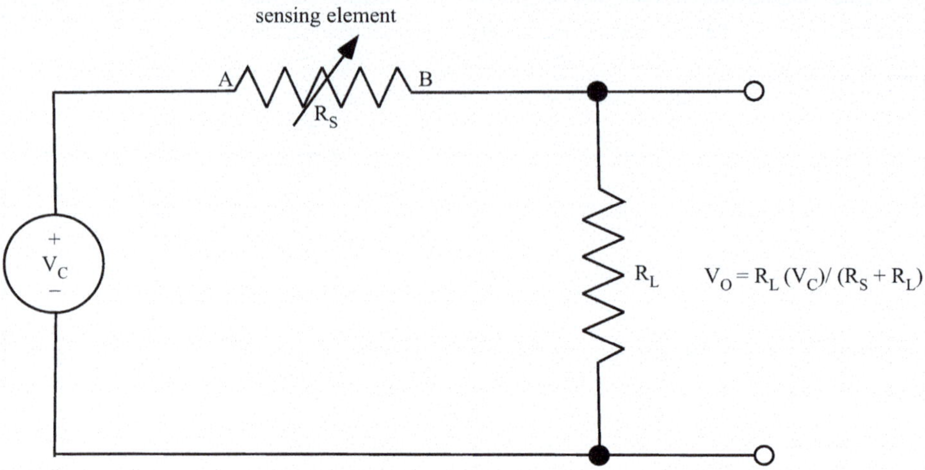

a) electrical resistance (RS) of the sensing element varies in the presence of a specific gas or gases.

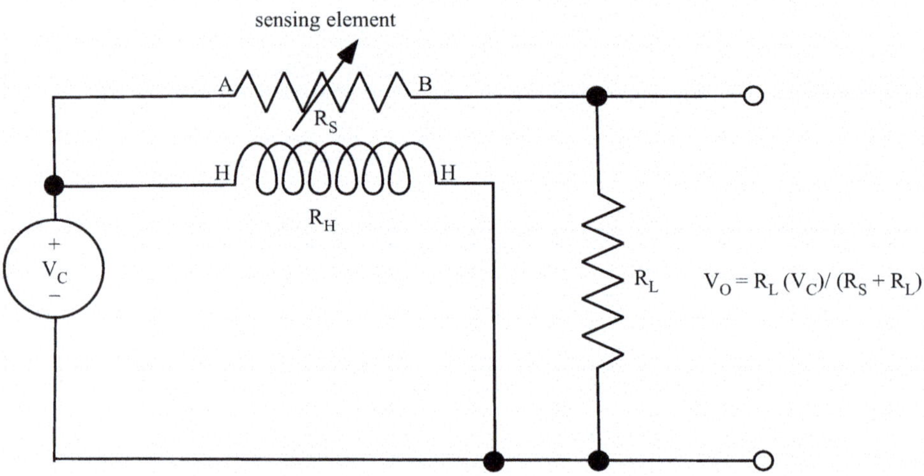

b) many of the MQ series of sensors require an applied heater voltage.

c) MQ sensors and MQ breakout board
(Sensor images courtesy of SparkFun Electronics, Inc. CC BY-NC-SA (www.sparkfun.com)).

Fig. 4.14 MQ sensor interface

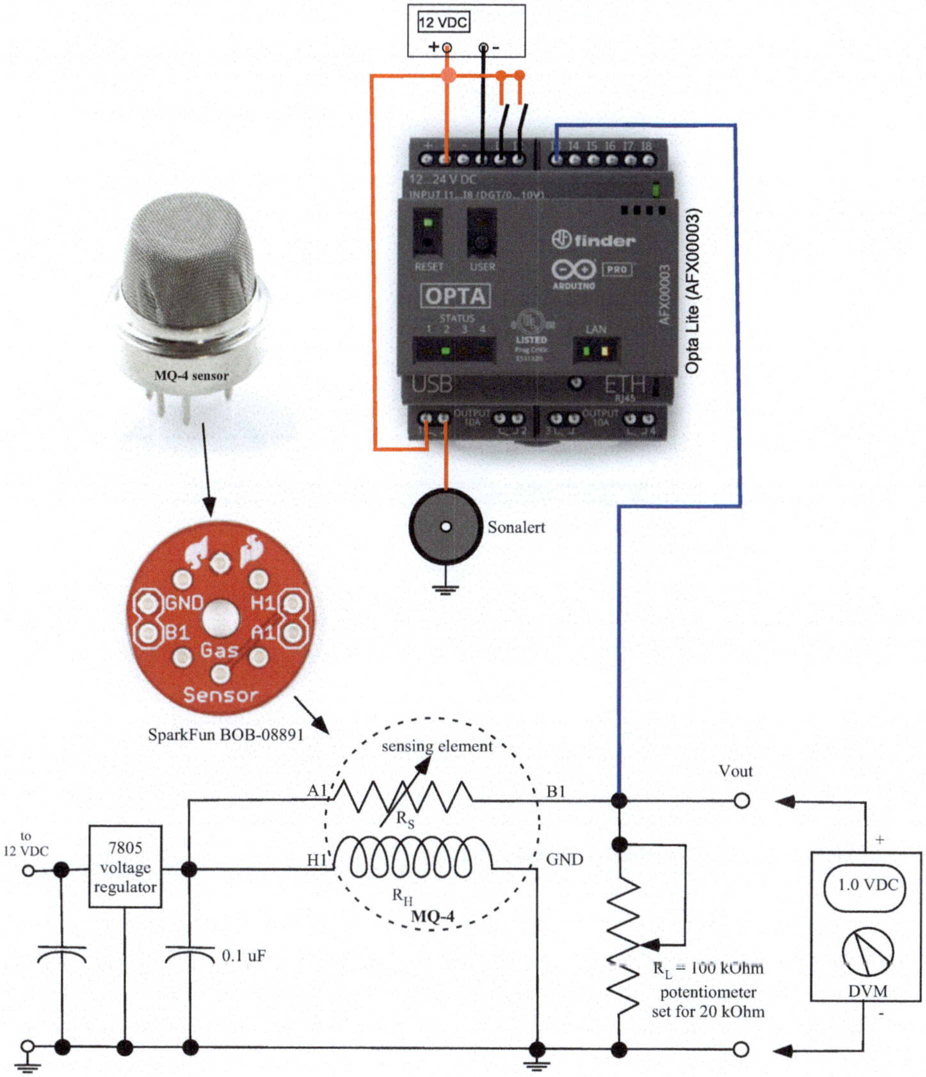

Fig. 4.15 MQ sensor test circuit. Sensor images courtesy of Sparkfun electronics, (CC BY–NC–SA) (www.sparkfun.com)

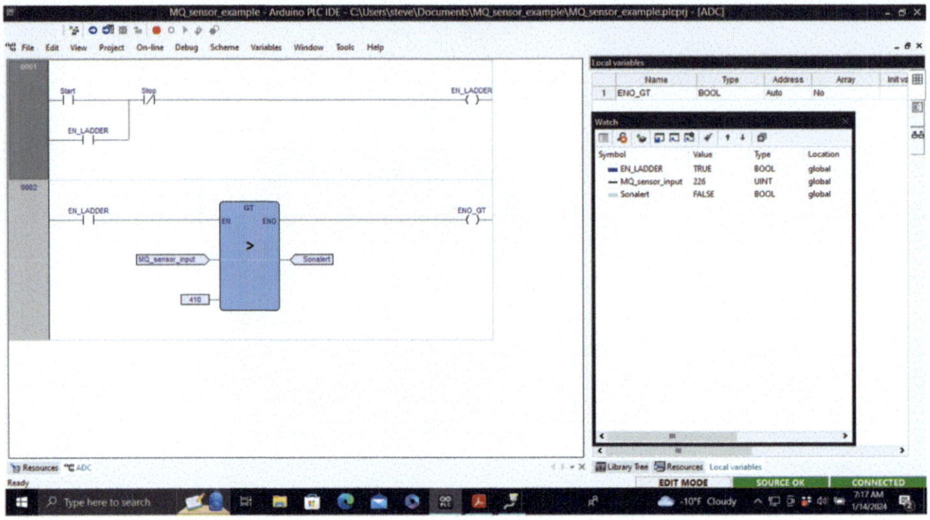

Fig. 4.16 MQ sensor ladder logic sketch

4.4.7 Greenhouse Sensors

In Chap. 5 we explore a variety of sensors to measure the temperature, humidity, soil moisture, and rain barrel water level within an instrumented greenhouse.

4.5 Output Devices and Actuators

An external device should not be connected to a controller without first performing careful interface analysis to ensure the voltage, current, and timing requirements of the controller and the external device are compatible. In this section, we describe interface considerations for a wide variety of external devices. We begin with the interface for a single light emitting diode.

4.6 Light Emitting Diodes (LEDs)

An LED is typically used as a logic or status indicator to inform the presence of a logic one or a logic zero at a specific pin of a microcontroller. An LED has two leads: the anode or positive lead and the cathode or negative lead.

To properly bias an LED, the anode lead must be biased at a level approximately 1.7–2.2 V higher than the cathode lead. This specification is known as the forward voltage (V_f) of the

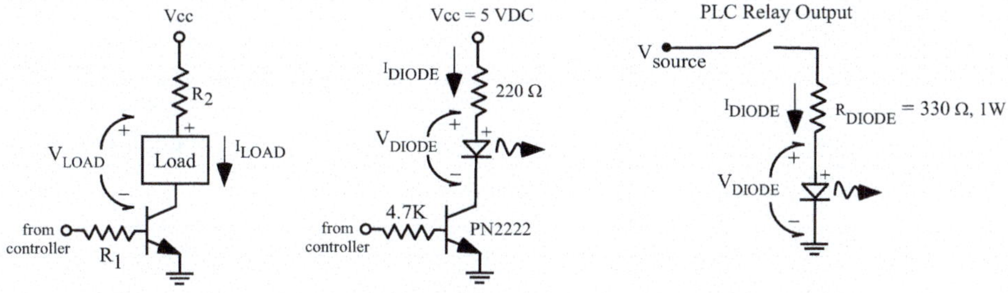

a) Interface circuit.

b) LED interface (V_{DIODE} = 2 V, I_{DIODE} = 15 mA)

c) LED interface for PLC (V_{DIODE} = 2 V, I_{DIODE} = 30 mA)

Fig. 4.17 LED interface

LED. The LED current must also be limited to a safe level known as the forward current (I_f). The diode voltage and current specifications are usually provided by the manufacturer.

A generic interface circuit for a controller is provided in Fig. 4.17a. A logic high control signal from a controller is applied to the transistor base. The NPN transistor acts as a switch and allows collector current flow from the voltage source (Vcc) through the load to ground. The value of the limiting resistor R_2 is calculated based on the desired load voltage and current. The resistor R_2 is calculated as:

$$R_2 = (V_{cc} - V_{LOAD})/I_{LOAD}$$

The value of R_1 is determined to ensure VBE is greater than 0.7 VDC. An example of an LED biasing circuit is provided in Fig. 4.17b. An NPN transistor such as a 2N2222 (PN2222 or MPQ2222) is used. A resistor value of 220 Ω is calculated for R_2.

Figure 4.17c provides the PLC interface circuit for a 10 mm LED with the voltage and current characteristics shown. The source voltage is set for 12 VDC.

4.7 Annunciators–Sonalerts, Beepers, Buzzers

In Fig. 4.15, we use a Sonalert (Mallory #PK–27N25WQ, 3–28 VDC) to provide an audible warning of detected smoke. The Sonalert may be directly driven by the Opta PLC relay output. Sonalerts, beepers, and buzzers are available in a wide variety of audible frequencies.

4.8 Electromechanical Devices

In this section we discuss the interface of a number of electromechanical devices including DC motors, linear actuators, pumps, and solenoid controlled valves.

4.9 DC Motors

Often a controller is used to control a high power motor load. To properly interface the motor to the controller, we must be familiar with the different types of motor technologies. Motor types are illustrated in Fig. 4.18.

- **DC motor**: A DC motor has a positive and negative terminal. When a DC power supply of suitable current rating is applied to the motor it will rotate. If the polarity of the supply is switched with reference to the motor terminals, the motor will rotate in the opposite direction. The speed of the motor is roughly proportional to the applied voltage up to the rated voltage of the motor.
- **Servo motor**: A servo motor provides a precision angular rotation for an applied pulse width modulation duty cycle. As the duty cycle of the applied signal is varied, the angular displacement of the motor also varies. This type of motor is used to change mechanical positions such as the steering angle of a wheel.
- **Stepper motor**: A stepper motor as its name implies provides an incremental step change in rotation (typically 2.5° per step) for a step change in control signal sequence. The motor is typically controlled by a two or four wire interface. For the four wire stepper motor, the microcontroller provides a four bit control sequence to rotate the motor clockwise. To turn the motor counterclockwise, the control sequence is reversed. For low power control signals, MOSFETs or power transistors are used to provide for the proper voltage and current requirements of the pulse sequence. Many PLCs have the capability to drive the stepper motor directly.

Space does not allow a full discussion of all motor types. We will concentrate on those common in industrial control applications. Let's take a closer look at several motor types.

4.10 DC Motor Speed and Direction Control

As shown in Fig. 4.18a, the speed of a DC motor is varied by changing the effective voltage delivered. At the rated motor voltage, the motor runs a full speed. When fifty percent of the rated motor voltage is delivered to the motor, it runs at approximately 50% of rated speed, etc. The effective voltage delivered to the motor is controlled by pulse width modulation (PWM).

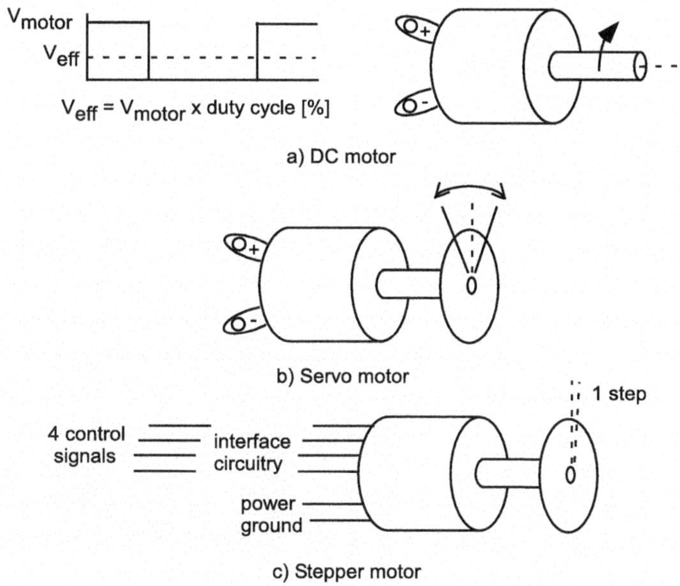

Fig. 4.18 Motor types

4.10.1 Pulse Width Modulation

Motor speed may be varied by changing the applied motor voltage. PWM control signal techniques may be used to precisely control the motor speed. With PWM the duty cycle of the motor control signal is varied.

The duty cycle of the PWM signal expressed as a percentage (high time/period * 100) will also be the percentage of the motor supply voltage applied to the motor, and hence the percentage of rated full speed at which the motor will rotate. We explore DC motor control techniques in the Application section later in the chapter.

Example. This is provided as an illustrative example. It is not practical for a real world application due to the mechanical characteristics of the Opta's mechanical relay outputs. With that said, there are important concepts here to explore.

The baseline Opta PLC is not equipped with a dedicated hardware based pulse width modulation system[5]. Instead, we use a technique called "bit banging" to emulate a PWM system using software. Bit banging may be employed in many applications when a software emulation replaces a dedicated hardware system (e.g. UART, SPI, etc.).

A Brother 12 VDC, 1,500 RPM DC motor equipped with an optical tachometer is used. The hardware configuration for the example is provided in Fig. 4.19.

[5] In Chap. 6, we explore Opta expansion modules that provide dedicated PWM support.

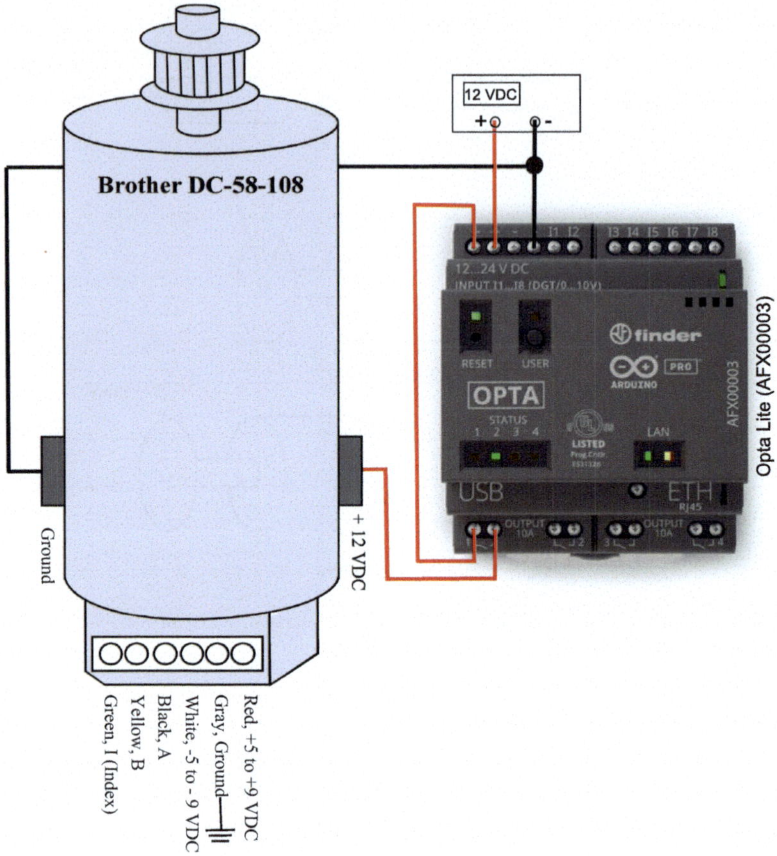

Fig. 4.19 Opta DC motor demonstration circuit. Images used courtesy of the Arduino team (CC BY–NC–SA) (www.arduino.cc)

```
//*****************************************************************************
//Opta_motor_demo
//- 12 VDC, 1500 RPM motor
//- Demonstrates 'bit bang' PWM
//- PWM duty cycle provides motor 8 VDC and ~1000 RPM
//
//S. Barrett, 01.15.24
//******************************************************************************

int          i;
int          motor_high_time, motor_low_time;
int          troubleshoot = 0;              //turns serial monitor on/off

void setup()
{
Serial.begin(9600);                         //serial monitor BAUD rate
pinMode(D0, OUTPUT);                        //Initialize Relays outputs
```

```
pinMode(LED_D0, OUTPUT);                          //Initialize Opta LEDs

digitalWrite(D0, HIGH);                           //Relay 1 on
digitalWrite(LED_D0, HIGH);                       //LED 1 on
delay(motor_high_time);                           //PWM: high baseline

digitalWrite(D0, LOW);                            //Relay 1 off
digitalWrite(LED_D0, LOW);                        //LED 1 off
delay(motor_low_time);                            //PWM: low baseline
}

void loop()
{
Serial.println("Duty cycle:25%");
for(i = 0; i <= 99; i++)                           //duty cycle: 25%
  {
  digitalWrite(D0, HIGH);                         //Relay 1 on
  digitalWrite(LED_D0, HIGH);                     //LED 1 on
  delay(25);                                     //PWM: high baseline

  digitalWrite(D0, LOW);                          //Relay 1 off
  digitalWrite(LED_D0, LOW);                      //LED 1 off
  delay(75);                                     //PWM: low baseline
  }

Serial.println("Duty cycle:50%");
for(i = 0; i <= 99; i++)                           //duty cycle: 50%
  {
  digitalWrite(D0, HIGH);                         //Relay 1 on
  digitalWrite(LED_D0, HIGH);                     //LED 1 on
  delay(50);                                     //PWM: high baseline

  digitalWrite(D0, LOW);                          //Relay 1 off
  digitalWrite(LED_D0, LOW);                      //LED 1 off
  delay(50);                                     //PWM: low baseline
  }

Serial.println("Duty cycle:75%");
for(i = 0; i <= 99; i++)                           //duty cycle: 75%
  {
  digitalWrite(D0, HIGH);                         //Relay 1 on
  digitalWrite(LED_D0, HIGH);                     //LED 1 on
  delay(75);                                     //PWM: high baseline

  digitalWrite(D0, LOW);                          //Relay 1 off
  digitalWrite(LED_D0, LOW);                      //LED 1 off
  delay(25);                                     //PWM: low baseline
  }

}

//******************************************************************
```

4.10.2 H Bridge Direction Control

For a DC motor to operate in both the clockwise and counter clockwise direction, the polarity of the DC motor supplied must be changed. To operate the motor in the forward direction, the positive battery terminal must be connected to the positive motor terminal while the negative battery terminal must be provided to the negative motor terminal. To reverse the motor direction the motor supply polarity must be reversed.

An H–bridge is a circuit employed to perform this polarity switch. The H–bridge circuit consists of four electronic switches as shown In Fig. 4.20. For forward motor direction switches 1 and 4 are closed; whereas, for reverse direction switches 2 and 3 are closed.

Low power H–bridges (500 mA) come in a convenient dual in line package (e.g., 754110). For higher power motors, a H–bridge may be constructed from discrete components as shown in Fig. 4.20. The ZTX451 and ZTX551 are NPN and PNP transistors with similar characteristics. The 11DQ06 are Schottky diodes. For driving higher power loads, the switching devices are sized appropriately.

If PWM signals are used to drive the base of the transistors (from microcontroller pins pin 2 and pin 3), both motor speed and direction may be controlled by the circuit. The transistors

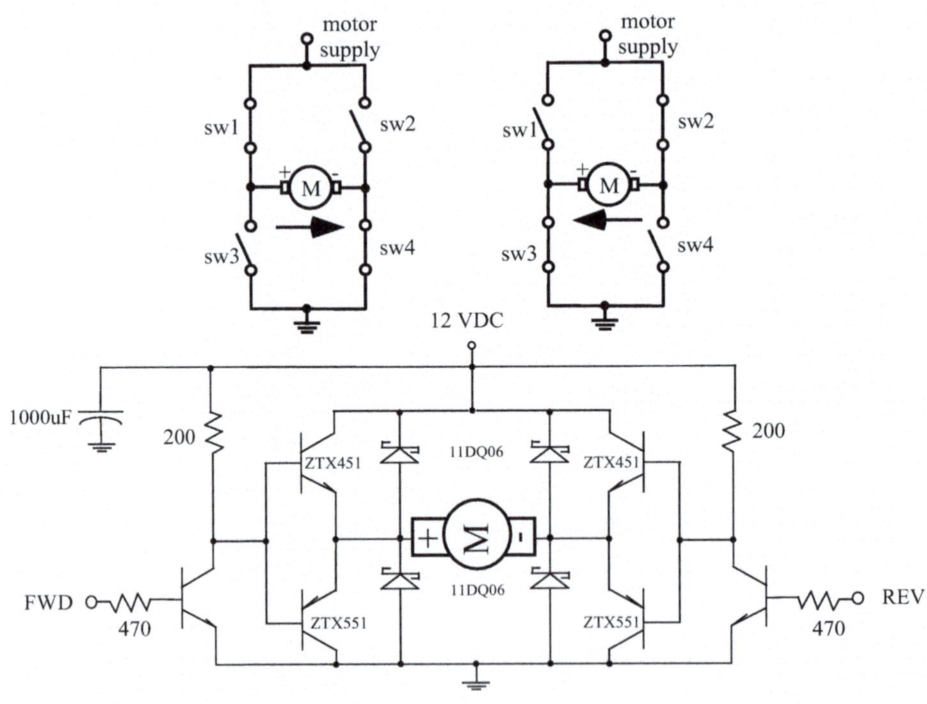

Fig. 4.20 H–bridge control circuit

used in the circuit must have a current rating sufficient to handle the current requirements of the motor during start and stall conditions.

4.11 Linear Actuator

A linear actuator is a specially designed motor that converts rotary to linear motion. The linear actuator is equipped with a mechanical rod that is extended when asserted in one direction and retracted when the polarity of assertion is reversed. An H–bridge may be used to control a linear actuator as shown in Fig. 4.21. In this circuit the Opta provides forward and reverse signals for the H–bridge in response to user input. Linear actuator forward and reverse switches are provided at Opta inputs A0 and A1.

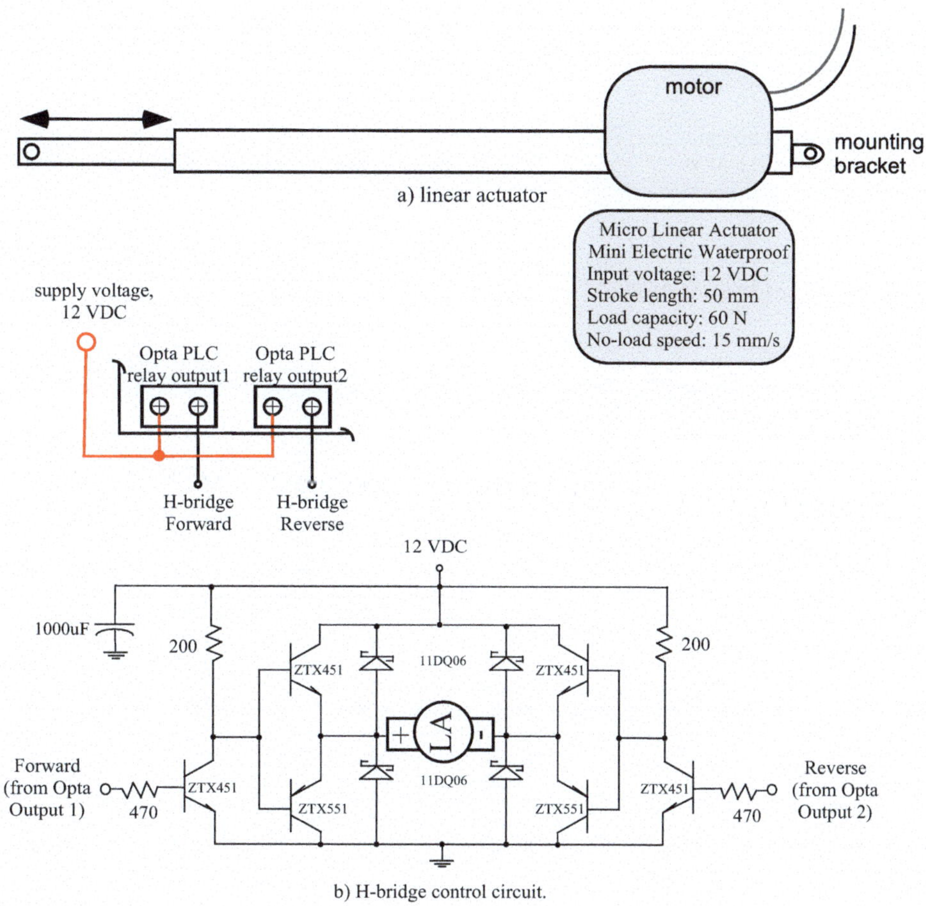

Fig. 4.21 Linear actuator control circuit (O'Berto)

```
//***************************************************************
//Opta_linear_actuator2
//***************************************************************

void setup()
{
pinMode(PIN_A0, INPUT);          //Opta I1: FWD
pinMode(PIN_A1, INPUT);          //Opta I2: REV

pinMode(LED_D0, OUTPUT);         //Opta LED1: FWD_LED
pinMode(LED_D1, OUTPUT);         //Opta LED2: REV_LED

pinMode(D0, OUTPUT);             //Opta D0: FWD_OUT
pinMode(D1, OUTPUT);             //Opta D1: REV_OUT
}

void loop()
{
if(digitalRead(PIN_A0))
  {
  digitalWrite(LED_D0, HIGH);
  digitalWrite(D0, HIGH);
  delay(100);
  digitalWrite(LED_D0, LOW);
  digitalWrite(D1, LOW);
  delay(100);
  }
else if(digitalRead(PIN_A1))
  {
  digitalWrite(LED_D1, HIGH);
  digitalWrite(D1, HIGH);
  delay(100);
  digitalWrite(LED_D1, LOW);
  digitalWrite(D1, LOW);
  delay(100);
  }
else
  {
  digitalWrite(LED_D0, LOW);
  digitalWrite(D0, LOW);
  digitalWrite(LED_D1, LOW);
  digitalWrite(D1, LOW);
  delay(100);
  }
}

//***************************************************************
```

4.12 Stepper Motor Control

Stepper motors are used to provide a discrete angular displacement in response to a control signal step. There are a wide variety of stepper motors including bipolar and unipolar types with different configurations of motor coil wiring. Due to space limitations we only discuss the unipolar, five wire stepper motor. The internal coil configuration for this motor is shown in Fig. 4.22b.

Often, a wiring diagram is not available for the stepper motor. Based on the wiring configuration (Reference Fig. 4.22b), one can find out the common line for both coils. It has a resistance that is one–half of all of the other coils. Once the common connection is found, one can connect the stepper motor into the interface circuit. By changing the other connections, one can determine the correct connections for the step sequence. To rotate the motor either clockwise or counterclockwise, a specific step sequence must be sent to the motor control wires as shown in Fig. 4.22c and f.

A microcontroller does not have sufficient capability to drive the motor directly. Therefore, an interface circuit is required as shown in Fig. 4.22d. The Opta PLC relay outputs have a sufficient current rating to drive the stepper motor directly as shown in (e). The speed of motor rotation is determined by how fast the control sequence is completed.

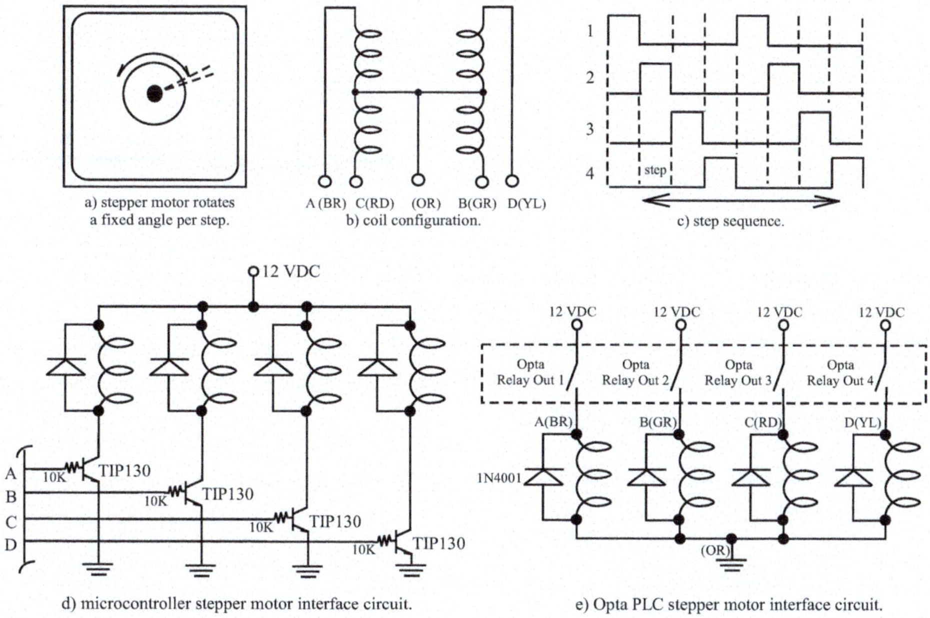

Fig. 4.22 Unipolar stepper motor control. Images used courtesy of the Arduino team (CC BY–NC–SA) (www.arduino.cc)

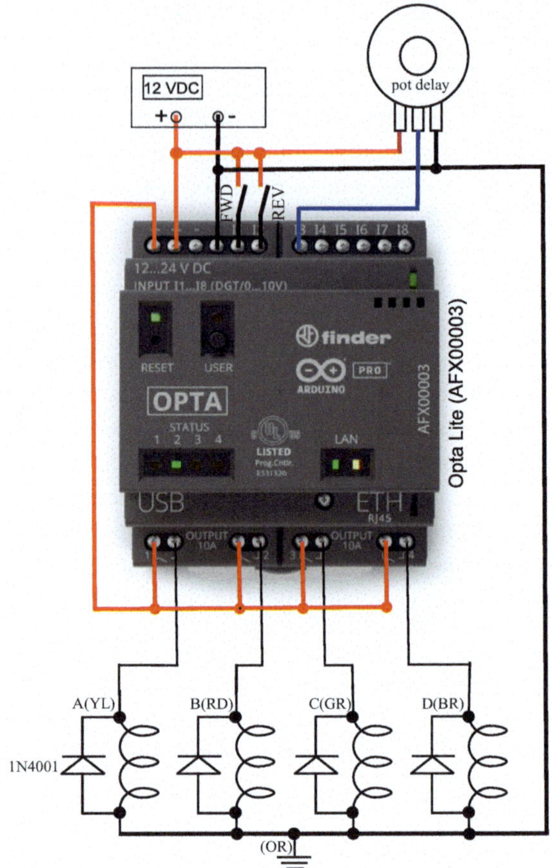

Fig. 4.23 Unipolar stepper motor control circuit

Example: The Opta PLC relay outputs have a sufficient current rating to drive the stepper motor directly. In the following example, we use the Opta PLC to control a five wire JRP 42BYG016 stepper motor. The motor requires 12 VDC, 160 mA per phase. As shown in Fig. 4.23 a pushbutton switch is used to determine motor direction (forward/reverse) and a potentiometer is used to set the step delay.

```
//*************************************************************
//Opta_stepper2
//
//Purpose: Test output relays of the Opta with stepper motor
//  - Use pushbutton (Opta I1) to set motor forward
//  - Use pushbutton (Opta I2) to set motor forward
//  - Use potentiometer (Opta I3) to set additional step
//    delay (pot_delay)
//
//author Arduino
//Modified: S. Barrett, 11.17.24
```

```
//
// This is Open Source software.
//**********************************************************

unsigned int step_delay = 100;
unsigned int pot_delay;
unsigned int last_step = 1;
unsigned int next_step;

void setup()
{
Serial.begin(9600);
analogReadResolution(12); //4095 is the max value with 12 bits
pinMode(PIN_A0, INPUT);   //Opta inputs I1: FWD
pinMode(PIN_A1, INPUT);   //Opta inputs I2: REV

pinMode(D0, OUTPUT);      //Initialize Relays outputs
pinMode(D1, OUTPUT);
pinMode(D2, OUTPUT);
pinMode(D3, OUTPUT);

pinMode(LED_D0, OUTPUT);  //Initialize Opta LEDs
pinMode(LED_D1, OUTPUT);
pinMode(LED_D2, OUTPUT);
pinMode(LED_D3, OUTPUT);
}

void loop()
{
pot_delay = analogRead(A2);
//Print out the value of pot_delay read from I3
Serial.print("Pot delay (ms): ");
Serial.println(pot_delay);

if(digitalRead(PIN_A0))           //Forward
  {
  while(digitalRead(PIN_A0))      //while FWD switch pressed
    {
    if(last_step == 1)
      {
      //Closes/opens contact relay 1 and turns on/off LED 1
      Serial.println("Forward - step 1");
      digitalWrite(D0, HIGH);         //Sets relay 1 on
      digitalWrite(LED_D0, HIGH);
      delay(step_delay + pot_delay);
      digitalWrite(D0, LOW);          //Sets relay 1 off
      digitalWrite(LED_D0, LOW);
      delay(step_delay + pot_delay);
      next_step = 2;
      }

    else if (last_step == 2)
      {
      //Closes/opens contact relay 2 and turns on/off LED 2
      Serial.println("Forward - step 2");
      digitalWrite(D1, HIGH);         //Sets relay 2 on
```

```
      digitalWrite(LED_D1, HIGH);
      delay(step_delay + pot_delay);
      digitalWrite(D1, LOW);          //Sets relay 2 off
      digitalWrite(LED_D1, LOW);
      delay(step_delay + pot_delay);
      next_step = 3;
      }

    else if (last_step == 3)
      {
      //Closes/opens contact relay 3 and turns on/off LED 3
      Serial.println("Forward - step 3");
      digitalWrite(D2, HIGH);         //Sets relay 3 on
      digitalWrite(LED_D2, HIGH);
      delay(step_delay + pot_delay);
      digitalWrite(D2, LOW);          //Sets relay 3 off
      digitalWrite(LED_D2, LOW);
      delay(step_delay + pot_delay);
      next_step = 4;
      }

    else if (last_step == 4)
      {
      //Closes/opens contact relay 4 and turns on/off LED 4
      Serial.println("Forward - step 4");
      digitalWrite(D3, HIGH);         //Sets relay 4 on
      digitalWrite(LED_D3, HIGH);
      delay(step_delay + pot_delay);
      digitalWrite(D3, LOW);          //Sets relay 4 off
      digitalWrite(LED_D3, LOW);
      delay(step_delay + pot_delay);
      next_step = 1;
      }
    else
      {
      ;
      }
    last_step = next_step;
    }//end while
  }//end if

if(digitalRead(PIN_A1))           //Reverse
  {
  while(digitalRead(PIN_A1))      //while REV switch pressed
    {
    if(last_step == 4)
      {
      //Closes/opens contact relay 4 and turns on/off LED 4
      Serial.println("Reverse - step 4");
      digitalWrite(D3, HIGH);         //Sets relay 4 on
      digitalWrite(LED_D3, HIGH);
      delay(step_delay + pot_delay);
      digitalWrite(D3, LOW);          //Sets relay 4 off
      digitalWrite(LED_D3, LOW);
      delay(step_delay + pot_delay);
      next_step = 3;
```

```
      }

    else if (last_step == 3)
      {
      //Closes/opens contact relay 3 and turns on/off LED 3
      Serial.println("Reverse - step 3");
      digitalWrite(D2, HIGH);         //Sets relay 3 on
      digitalWrite(LED_D2, HIGH);
      delay(step_delay + pot_delay);
      digitalWrite(D2, LOW);          //Sets relay 3 off
      digitalWrite(LED_D2, LOW);
      delay(step_delay + pot_delay);
      next_step = 2;
      }

    else if (last_step == 2)
      {
      //Closes/opens contact relay 2 and turns on/off LED 2
      Serial.println("Reverse - step 2");
      digitalWrite(D1, HIGH);         //Sets relay 2 on
      digitalWrite(LED_D1, HIGH);
      delay(step_delay + pot_delay);
      digitalWrite(D1, LOW);          //Sets relay 2 off
      digitalWrite(LED_D1, LOW);
      delay(step_delay + pot_delay);
      next_step = 1;
      }

    else if (last_step == 1)
      {
      //Closes/opens contact relay 1 and turns on/off LED 1}
      Serial.println("Reverse - step 1");
      digitalWrite(D0, HIGH);         //Sets relay 1 on
      digitalWrite(LED_D0, HIGH);
      delay(step_delay + pot_delay);
      digitalWrite(D0, LOW);          //Sets relay 1 off
      digitalWrite(LED_D0, LOW);
      delay(step_delay + pot_delay);
      next_step = 4;
      }
    else
      {
      ;
      }
    last_step = next_step;
    }//end while
  }//end if
}

//*****************************************************
```

4.12.1 Sequencer Control Logic

In many control applications, a sequencer is used to implement a repeatable fixed sequence of control signals as shown in Fig. 4.24a.

A sequencer is implemented using a series of binary counters operating at different clock frequencies. The output from the counters are provided to combinational logic circuitry to generate the desired control signal sequence as shown in Fig. 4.24b.

To implement a binary counter in ladder logic, cross coupled timers (e.g. TON) are used. The output of the configuration provides a square wave output as shown in Fig. 4.24c (BharadwajReddy).

4.12.2 Stepper Motor Control–Ladder Logic Sequencer

To implement a stepper motor controller in ladder logic, two counters (A_Light and B_Light) are used. The first counter operates at twice the frequency of the second counter. The two counters with their accompanying inverted outputs (A_Light' and B_Light') provide the output sequence shown in Fig. 4.25a.

To achieve the desired stepper motor control signal as shown in Fig. 4.25c, the outputs from the counters are combined using combinational logic as shown in Fig. 4.25b. The example implemented with the Arduino PLC IDE is shown in Figs. 4.26 and 4.27.

4.13 DC Solenoid Control

A solenoid provides a mechanical insertion (or extraction) when asserted. Often the solenoid is coupled with valves to control fluid flow. The solenoid may be directly driven by the Opta PLC relay outputs.

Example: Water valve control: Solenoid controlled water valves are available from Adafruit (www.adafruit.com). There are plastic (#997)and brass (#996) valves available. The plastic valve activates from 6 VDC at 160 mA to 12 VDC at 320 mA while the brass valve activates from 6 VDC at 1.6 A to 12 VDC at 3 A. An interface circuit for the plastic water solenoid valve is provided in Fig. 4.28.

4.14 Transducer Interface Design (TID)

A transducer is used to convert a physical variable such as temperature, pressure, or light intensity to a voltage for data collection and analysis by a controller. The controller accepts voltages between its power supply value and ground for analog–to–digital conversion. It is

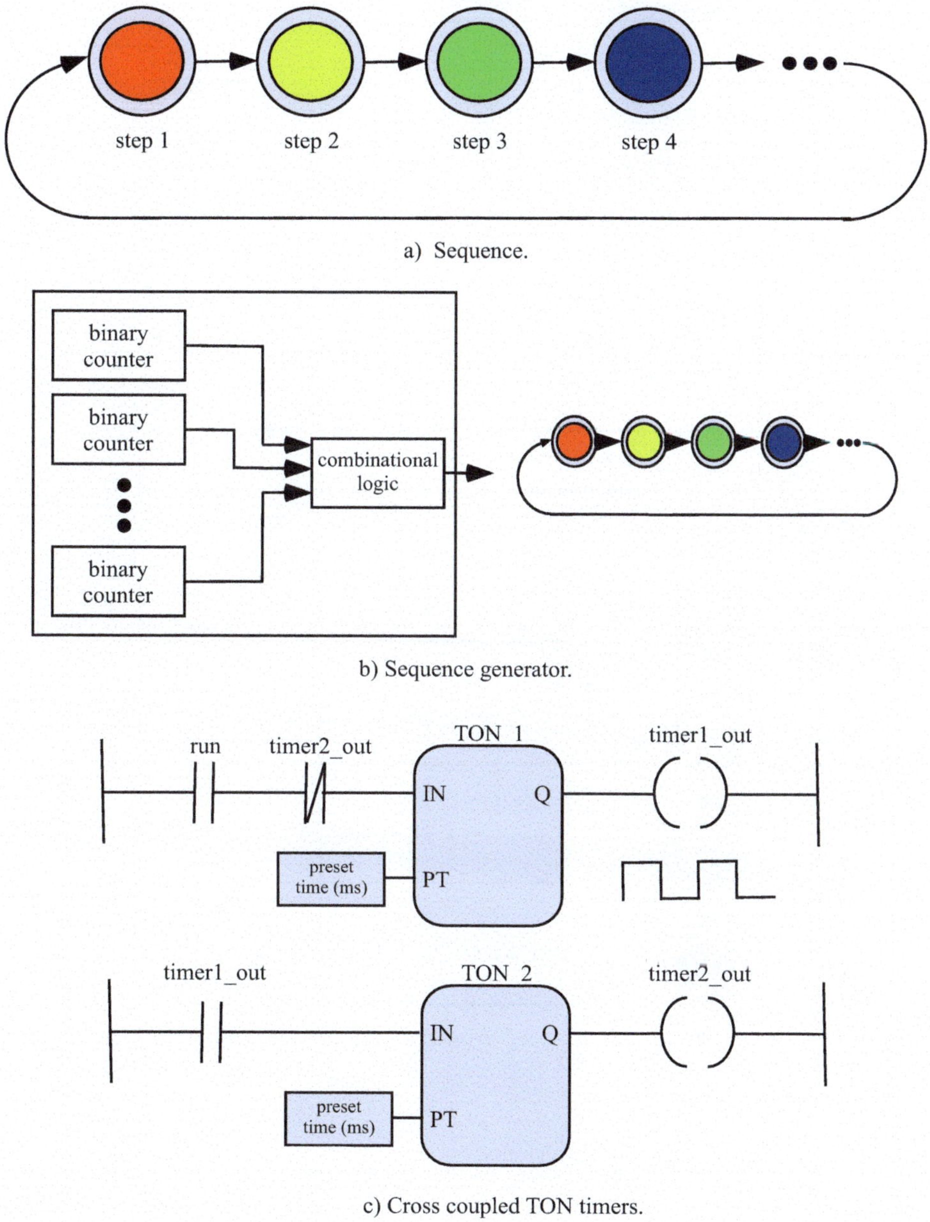

Fig. 4.24 Sequencer control logic

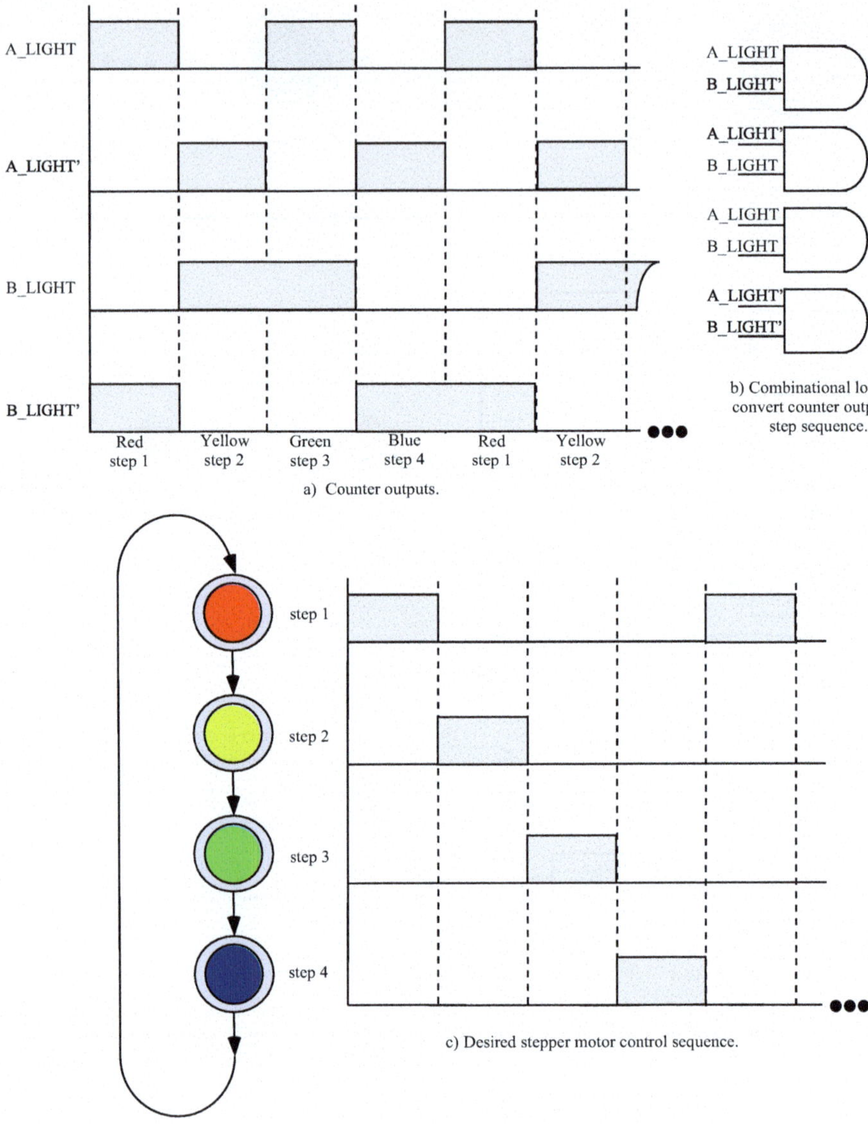

Fig. 4.25 Stepper motor controller in ladder logic

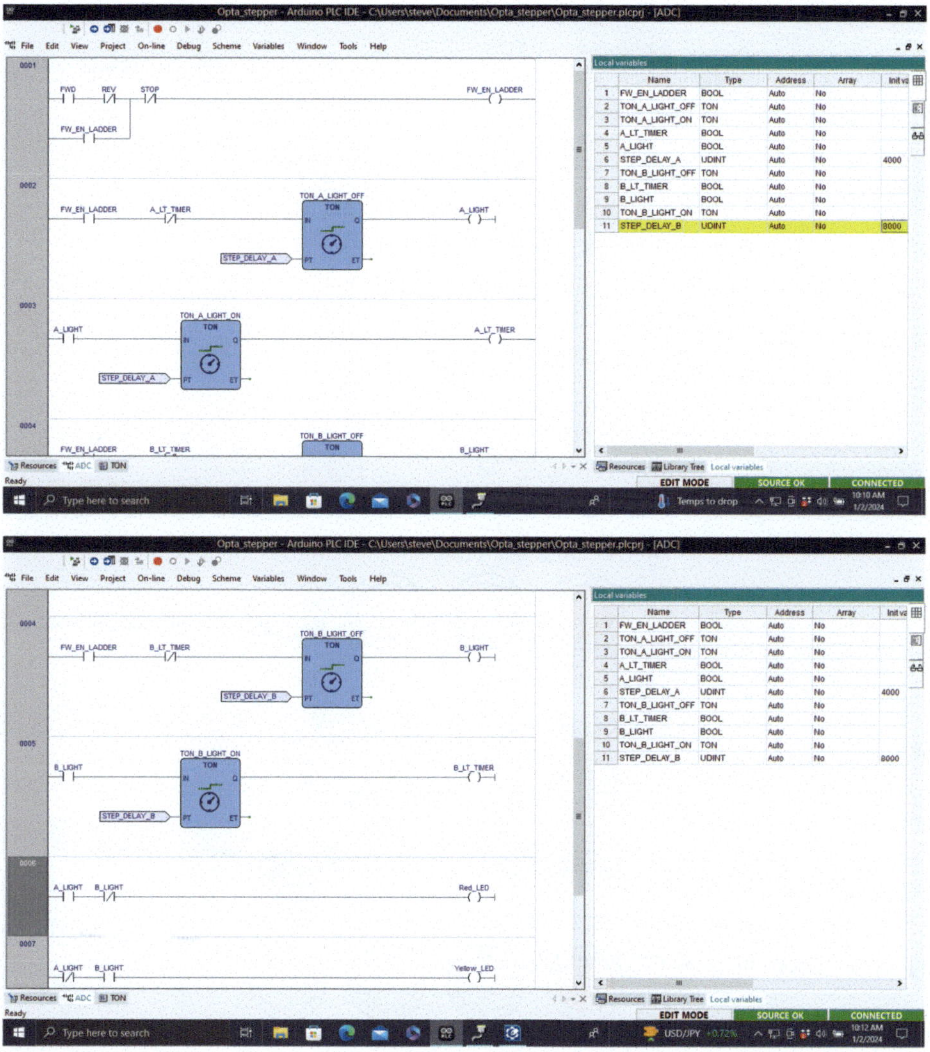

Fig. 4.26 Stepper motor controller in ladder logic. Using PLC IDE

the responsibility of the system designer to ensure transducer outputs are properly conditioned to meet these constraints.[6]

The signal conditioning circuitry is called the transducer interface. The objective of the transducer interface circuit is to scale and shift the electrical signal range to efficiently map the output of the input transducer to the input range of the analog–to–digital converter which

[6] The section on transducer interface design is adapted from "Electrical Signals and Systems," Department of Electrical Engineering, United States Air Force Academy.

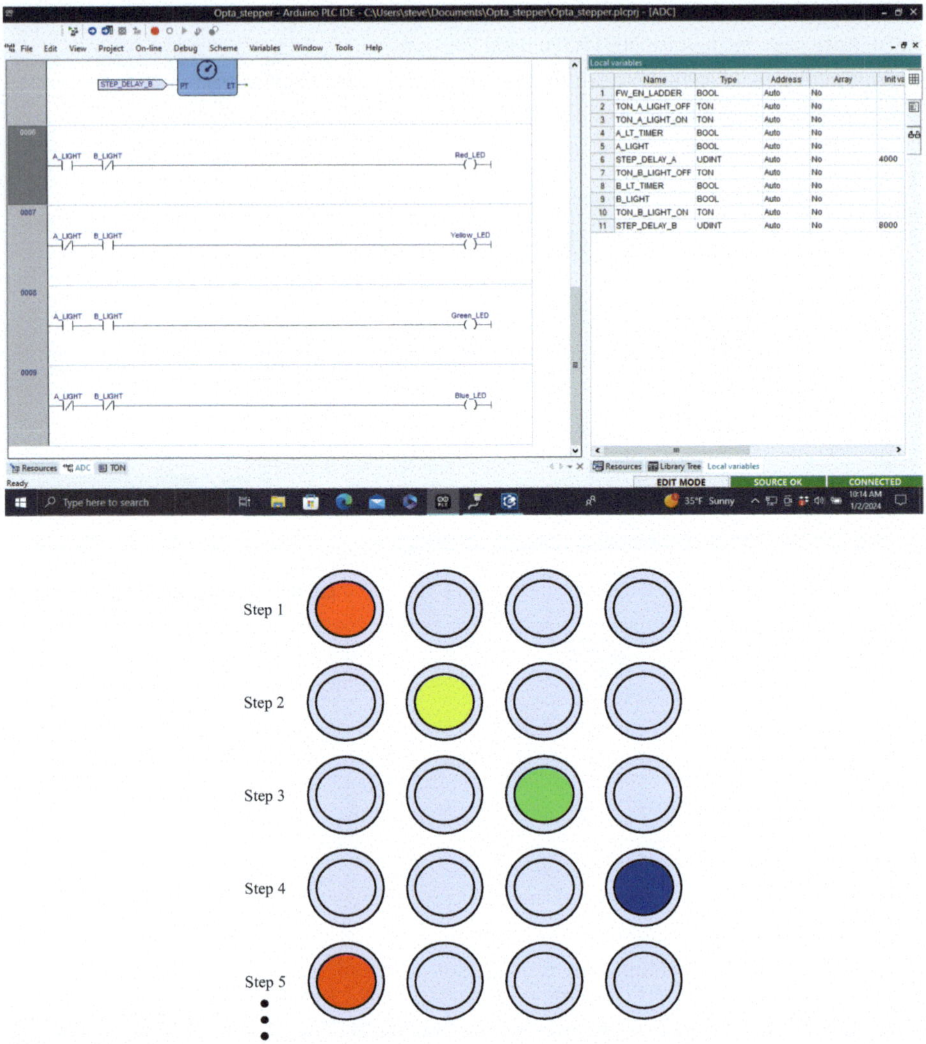

Fig. 4.27 Stepper motor controller in ladder logic. Using PLC IDE

is typically 0–5 VDC or 0–3.3 VDC for microcontrollers and 0–10 VDC from the Arduino Opta PLC.

Figure 4.29 shows the transducer interface circuit using an input transducer. This process assumes a linear input transducer. The output of the input transducer is first scaled by constant K. As an example, in the figure, we use a microphone as the input transducer whose output ranges from −5 VDC to +5 VDC. The input to the analog–to–digital converter ranges from 0 VDC to 5 VDC. The scalar multiplier with constant K maps the output range of the input

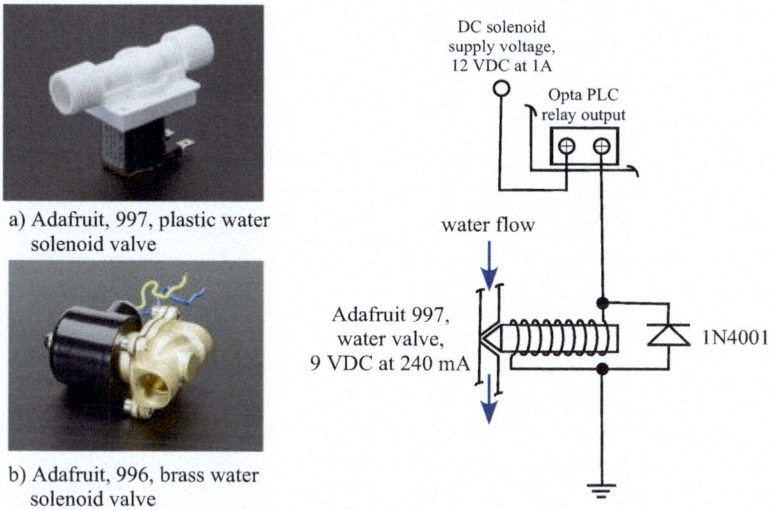

Fig. 4.28 Water valve interface circuit (www.adafruit.com). Illustration used with permission

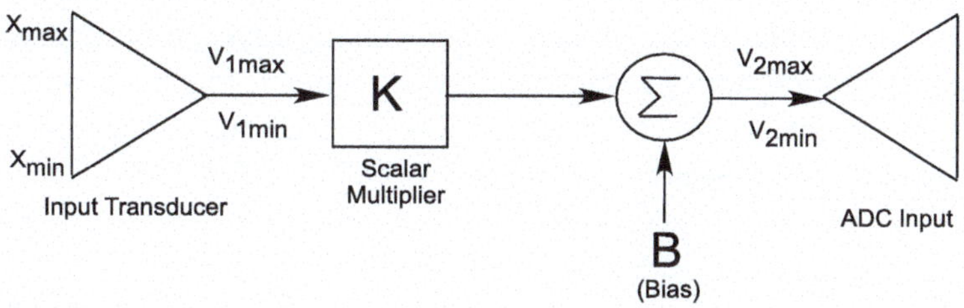

Fig. 4.29 Signal conditioning for ADC. A block diagram of the signal conditioning for an analog–to–digital converter. The range of the sensor voltage output is mapped to the analog–to–digital converter input voltage range. The scalar multiplier maps the magnitudes of the two ranges and the bias voltage is used to align two limits

transducer to the input range of the converter. Naturally, we need to multiply all input signals by $1/2$ to accommodate the mapping.

Once the range has been mapped, the signal now needs to be shifted. Note that the scale factor maps the output range of the input transducer as -2.5 VDC to $+2.5$ VDC instead of 0 VDC to 5 VDC. The second portion of the circuit, the bias stage, shifts the range by 2.5 VDC, thereby completing the correct mapping. Actual implementation of the circuit components are accomplished using operational amplifiers.

In general, the scaling and bias process may be described by two equations:

$$V_{2max} = (V_{1max} \times K) + B$$

$$V_{2min} = (V_{1min} \times K) + B$$

The variable $V_{1\max}$ represents the maximum output voltage from the input transducer. This voltage occurs when the maximum physical variable ($X_{\max}$) is presented to the input transducer. This voltage must be scaled by the scalar multiplier (K) and then a DC offset bias voltage (B) is added to provide the voltage $V_{2\max}$ to the input of the ADC converter.

Similarly, The variable $V_{1\min}$ represents the minimum output voltage from the input transducer. This voltage occurs when the minimum physical variable ($X_{\min}$) is presented to the input transducer. This voltage must be scaled by the scalar multiplier (K) and then have a DC offset bias voltage (B) added to produce voltage $V_{2\min}$ to the input of the ADC converter.

Usually, the values of $V_{1\max}$ and $V_{1\min}$ are provided with the documentation for the transducer. Also, the values of $V_{2\max}$ and V_{2min} are known. They are the high and low reference voltages for the ADC system (usually 5 and 0 VDC for a microcontroller). We thus have two equations and two unknowns to solve for K and B. The circuits to scale by K and add the offset B are usually implemented with operational amplifiers.

Example: A photodiode is a semiconductor device that provides an output current corresponding to the light impinging on its active surface. The photodiode is used with a transimpedance amplifier to convert the output current to an output voltage. A photodiode/transimpedance amplifier provides an output voltage of 0 V for maximum rated light intensity and −2.50 VDC output voltage for the minimum rated light intensity.

Calculate the required values of K and B for this light transducer so it may be interfaced to a microcontroller's ADC system.

$$V_{2\max} = (V_{1\max} \times K) + B$$

$$V_{2min} = (V_{1min} \times K) + B$$

$$5.0\,V = (0\,V \times K) + B$$

$$0\,V = (-2.50\,V \times K) + B$$

The values of K and B may then be determined to be 2 and 5 VDC, respectively. The transducer interface circuit is then implemented using operational amplifiers (op amps).

Example: It was determined that the values of K and B were 2 and 5 VDC, respectively. The two–stage op amp circuitry provided in Fig. 4.30 implements these values of K and B. The first stage provides an amplification of −2 due to the use of the inverting amplifier configuration. In the second stage, a summing amplifier is used to add the output of the first stage with a bias of −5 VDC. Since this stage also introduces a minus sign to the result, the overall result of a gain of 2 and a bias of +5 VDC is achieved.

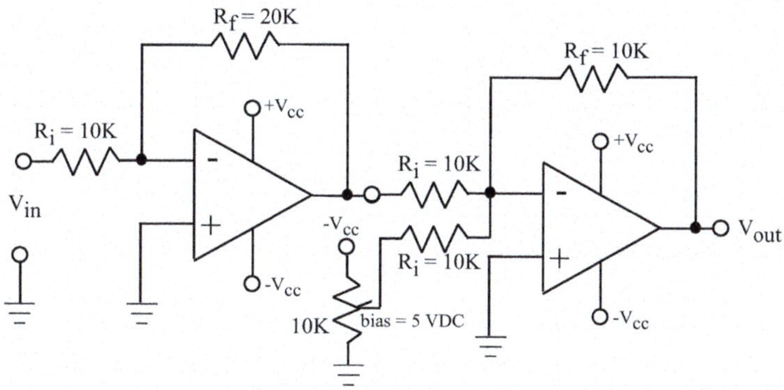

Fig. 4.30 Operational amplifier implementation of the transducer interface design (TID) example circuit

4.15 Operational Amplifier Overview

The operational amplifier or op amp is used extensively in applications to interface transducers. We begin the section exploring op amp origins and development. We then describe the ideal op amp and use it as a benchmark for real world, nonideal op amps. We investigate how to compensate for nonideal op amp parameters. Next, we review common op amp circuit configurations used extensively in instrumentation applications. We use the circuit configurations to explore transducer interface and applications.

4.15.1 Operational Amplifier Origins

The operational amplifier or op amp is a two input, single output amplifier. The output is an amplified version of the difference between the two inputs. It is quite common in bio-related instrumentation applications for signal amplification, conditioning, filtering, and may be used for mathematical operations.

The first op amps were developed in the 1940s using vacuum tube technology. In 1947 John Ragazzini in his paper "Analysis of Problems in Dynamics by Electronic Circuits" was the first to use the term operational amplifier. He and is co–authors wrote "The term 'operational amplifier' is a generic term applied to amplifiers whose gain functions are such as to enable them to perform certain useful operations such as summation, integration, differentiation, or a combination of such operations (Ragazzini)."

Op amp development continued for several decades with improvement in features and implementation technology. The first monolithic, i.e. single chip, integrated circuit op amp, the uA702, was designed by Bob Widlar of Fairchild Semiconductor. It was released in 1963. An improved op amp, the uA709, also designed by Widlar, was released in 1965. Widlar also designed the LM101 for the National Semiconductor Company. The uA741 op amp was released by Fairchild Semiconductor in 1968. It quickly became quite popular and is still in production today. An op amp with a single polarity power supply, the LM324, was released in 1972. It too is still quite popular today and is used in multiple applications (Jung).

4.15.2 Ideal Characteristics

A generic ideal operational amplifier is illustrated in Fig. 4.31. The op amp is an active device (requires power supplies) equipped with two inputs, a single output, and several DC voltage source inputs (Sedra and Smith, Faulkenberry).

The two op amp inputs are labeled Vp, or the noninverting input, and Vn, the inverting input. The output of the op amp is determined by taking the difference between Vp and Vn and multiplying the difference by the op amp's open loop gain (A_{vol}). This gain is typically a large value much greater than 50,000.

Due to the large value of A_{vol}, it does not take much of a difference between Vp and Vn before the op amp will saturate. When an op amp saturates, it does not damage the op amp, but the output is limited to values slightly less than the supply voltages $\pm V_{cc}$. This will clip the output, and hence distort the signal, at levels slightly less than $\pm V_{cc}$. To prevent saturation, op amps are typically used in a closed loop, negative feedback configuration.

Example: Comparator level detector. The comparator level detector is a common op amp building block. It is configured in an open loop configuration. One input is tied to a reference voltage threshold. The threshold setting is typically provided by a potentiometer connected between the supply voltage and ground. The input signal is provided to the other input as shown in Fig. 4.32 (Stout and Kaufman).

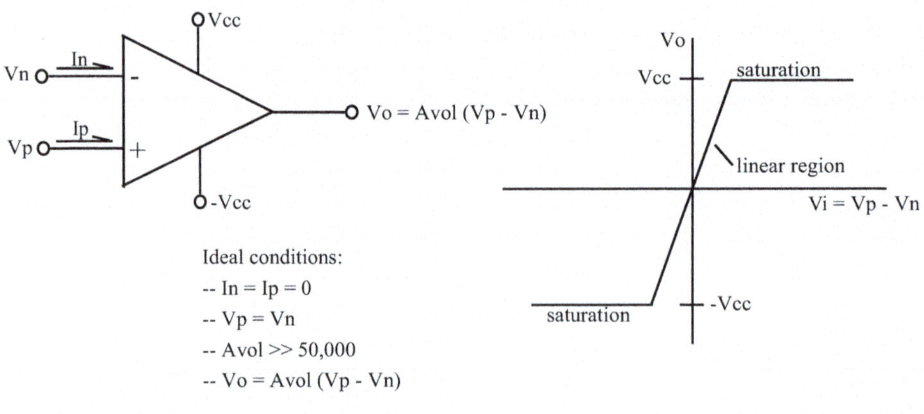

Fig. 4.31 Ideal operational amplifier characteristics

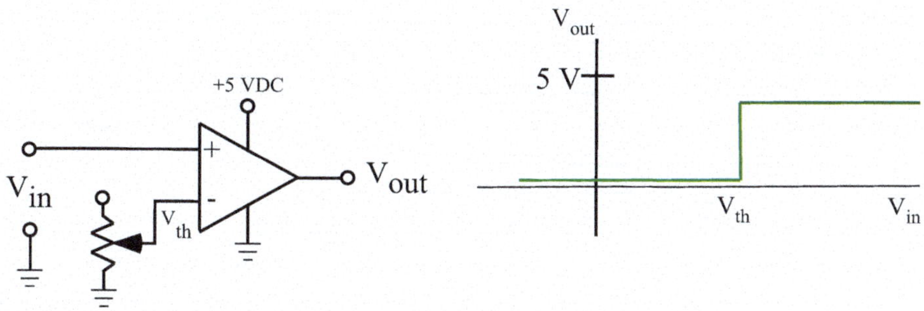

Fig. 4.32 Op amp comparator level detector circuit

When the input signal is higher than the threshold voltage, the op amp saturates toward the positive supply value. When the input signal is less than threshold signal, the op amp saturates toward the negative supply value. A comparator circuit may be used to restore a degraded digital signal to its original values. In this case a single–sided op amp such as the LM324 may be used with supply voltages of 5 VDC and ground for a 5 VDC digital system.

An ideal operational does not exist in the real world. However, it is a good first approximation for use in developing op amp application circuits. As shown in Fig. 4.31 an op amp has the following ideal characteristics (Sedra and Smith):

- Input currents In and Ip equal to zero;
- Infinite input impedance;
- Vp and Vn input voltages equal to one another;
- Extremely high open loop gain;
- Output impedance of zero;
- Infinite bandwidth;
- High slew rate; and
- Infinite common mode rejection.

We use these ideal conditions as a close approximation when developing characteristic equations for different op amp configurations.

4.15.3 Nonideal Characteristics

Ideal op amps do not exist in the real world although some come close. To better understand nonideal op amp features, we explore their origins and potential compensation methods (Sedra and Smith, Faulkenberry, Stout and Kaufman).

- **Frequency Response**: Ideally we desire an infinite bandwidth or frequency response. That is, we want the op amp to provide the same amplification response across the frequency spectrum. Due to internal capacitance and internal configurations, the frequency response usually has a lower and upper 3 dB point. A 3 dB point is a frequency where the op amp gain is down 3 dB from its passband value. In a specific application an op amp should be chosen that operates at the desired frequencies.
- **Gain Bandwidth Product**: The Gain Bandwidth Product (GBP) is a metric that describes the tradeoff between op amp voltage gain and desired frequency of operation. The GBP is a fixed value parameter for a specific op amp. Therefore, as the desired gain is increased the corresponding bandwidth decreases and vice versa. To provide a reasonable gain and bandwidth combination, the op amp is typically used in closed loop configurations as described in the next section.
- **Offset Voltage**: Typically an op amp consists of multiple stages. The first stage consists of a differential transistor pair with matched characteristics for the positive and negative portion of the amplifier. Any mismatch between the two input transistor characteristics is amplified. This leads to an op amp output voltage even when both inputs are at zero volts. Some op amps are equipped with offset null compensation inputs. A potentiometer may be connected between these inputs to provide an offset compensation for the mismatched inputs.
- **Bias Current**: The differential amplifier input described above requires small bias currents to maintain the input transistors in an active state. If each of the input bias currents flow through the same equivalent resistance, they are canceled out by the differential input amplifier configuration. A compensation resistor, $R_p = R_i || R_f$, may be connected between the positive input lead V_p and ground to minimize this effect. The resistors R_i and R_f are the input and feedback resistors.
- **Input resistance**: Ideally we want the input resistance R_i to be very high. This prevents loading down a previous op amp stage or having the current stage affect another stage's operation. This effect can be minimized by the wise choice of R_i values and using a voltage follower configuration between stages. The voltage follower provides no gain but provides high input impedance. We explore the voltage follower configuration in the next section.
- **Common Mode Rejection Ratio (CMRR)**: Ideally the op amp should amplify the difference between its two inputs while cancelling any voltages common to both inputs. This allows for the cancellation of noise common to both inputs. The Common Mode Rejection Ratio or CMRR is a metric comparing an op amp's differential gain (A_d) to its common mode gain (A_{cm}). It is expressed as:

$$CMRR = 20\,log(|A_d|/|A_{CM}|)$$

- **Slew Rate**: Ideally the op amp should amplify a signal with high fidelity even for high output amplitudes rapidly changing at high frequencies. The slew rate is a parameter provided in volts per microsecond describing the op amp's capability to do this.

4.15.4 Configurations

As described in previous sections, the op amp has a very large open loop gain which minimizes bandwidth and also the differential voltage applied to the inputs. Therefore, op amps are typically used in a closed loop configuration with a controlled gain to perform a variety of functions. A sample of classic operational amplifier configurations are provided in Fig. 4.33 (Faulkenberry).

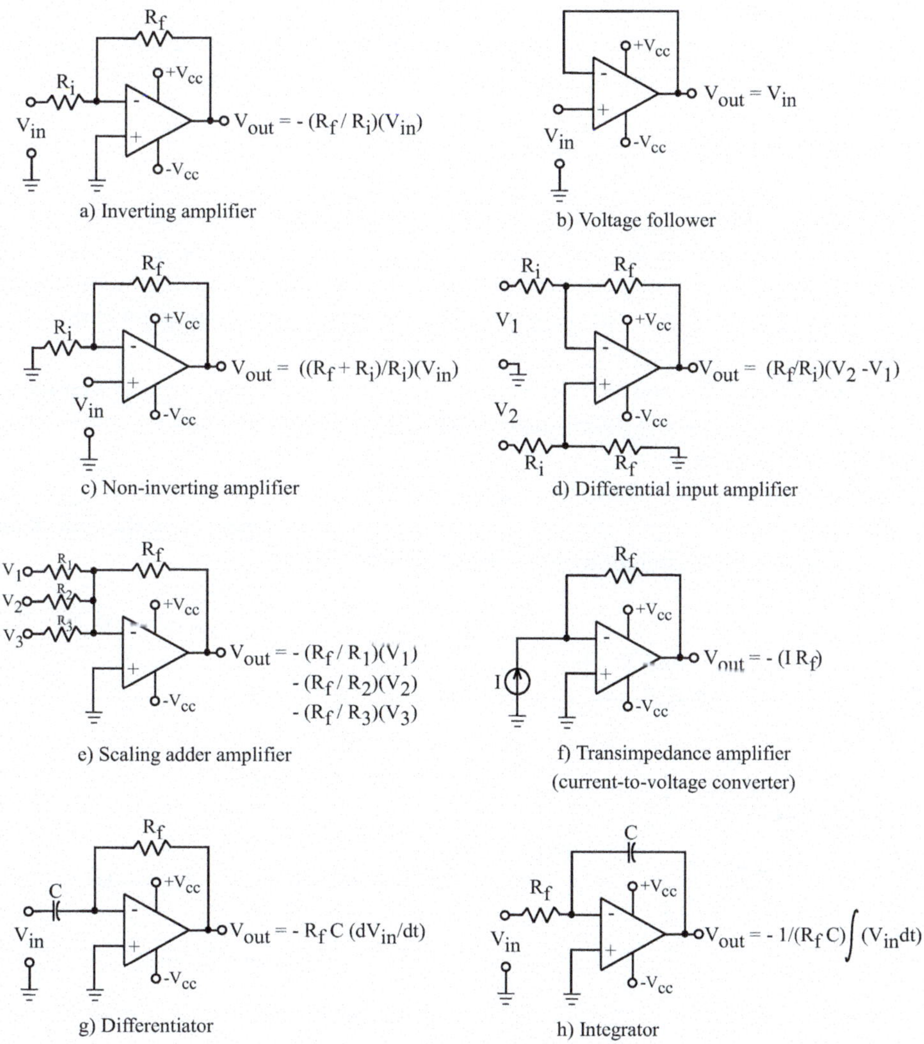

Fig. 4.33 Classic operational amplifier configurations. Adapted from [Faulkenberry]

It should be emphasized that the equations provided with each operational amplifier circuit are only valid if the circuit configurations are identical to those shown. Even a slight variation in the circuit configuration may have a dramatic effect on circuit operation.

It is important to analyze each operational amplifier circuit using the following steps:

- Write the node equation at Vn for the circuit.
- Apply ideal op amp characteristics to the node equation.
- Solve the node equation for Vo.

As an example, we provide the analysis of the noninverting amplifier circuit in Fig. 4.34. This same analysis technique may be applied to all of the circuits in Fig. 4.33 to arrive at the equations for Vout provided.

A brief description of each configurations follows.

- **Inverting amplifier**: The inverting amplifier provides a gain determined by $A_v = -R_f/R_i$. As indicated by the minus sign, the amplifier inverts the polarity of the input signal to produce the output signal. The value of R_i should be kept high to maintain a high input impedance. Also, the gain should be limited to prevent saturation of the output signal.
- **Voltage follower**: The voltage follower circuit provides a high impedance buffer for use between op amp stages in multi–stage designs. As the name implies, the output signal follows the input signal.
- **Noninverting amplifier**: The noninverting amplifier provides a noninverted gain. The value of R_i should be kept high to maintain a high input impedance. Also, the gain should be limited to prevent saturation of the output signal.

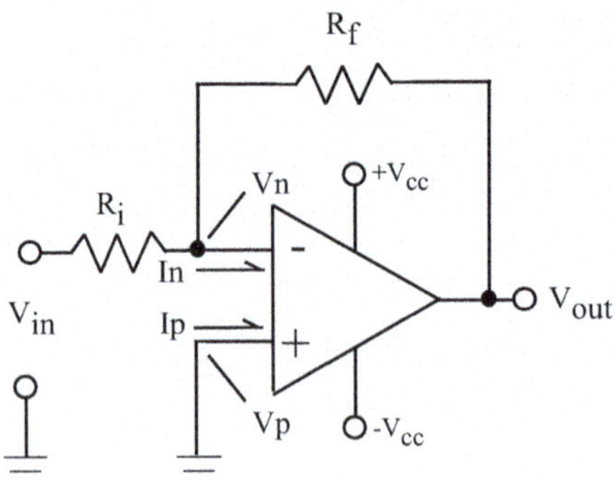

Node equation at Vn:

(Vn - Vin)/ Ri + (Vn - Vout)/Rf + In = 0

Apply ideal conditions:

In = Ip = 0

Vn = Vp = 0 (since Vp is grounded)

Solve node equation for Vout:

$V_{out} = -(R_f / R_i)(V_{in})$

Fig. 4.34 Operational amplifier analysis for the non–inverting amplifier. Adapted from (Faulkenberry)

- **Difference amplifier**: The difference amplifier provides the amplified difference of its two input signals. Voltages common to both inputs are not amplified (e.g. noise).
- **Summing amplifier**: The summing amplifier provides the amplified sum of the input signals. The input values of R_i may be chosen to determine the relative proportions of each input signal within the output signal. Also note the signal inversion at the output. The individual value of the input resistances (R_1, R_2, R_3) should be kept high to maintain high input impedance.
- **Transimpedance amplifier**: The transimpedance amplifier translates a current input to a voltage output. This configuration is commonly used to convert the current output from certain transduces to a voltage suitable for conversion by a microcontroller.
- **Integrator**: The integrator performs the mathematical integration of the input signal.
- **Differentiator**: The differentiator performs the mathematical integration of the input signal.

TMP36 Interface. An Analog Devices TMP36 low voltage temperature sensor is used to measure the interior temperature of a greenhouse. The TMP36 will be interfaced to the Opta PLC ADC using a transducer interface circuit. The TMP36 is a linear sensor providing a 10 mV per degree Centigrade output and measures temperatures between −40 to +125 °C. It provides 750 mV of output at 25 °C. www.analog.com.

To design the interface circuit, the minimum and maximum value of the temperature variable and corresponding output voltage must be known. A spreadsheet is provided in Fig. 4.35 to determine these values. Using two equations developed from the transducer interface design process, the values of K and B are determined to be 6.06 V/V and −0.606 V respectively. A block diagram and corresponding circuit diagram is provided in Fig. 4.36.

4.16 Application: DC Motor Speed Control

The goal of this example is to stabilize the speed of a DC motor using several different concepts discussed in this chapter. A block diagram of a circuit to stabilize motor speed is shown in Fig. 4.37.

The control algorithm is hosted on an Opta PLC. It takes as input the desired motor speed. The algorithm provides a pulse width modulated (PWM) signal to control motor speed. Motor speed is measured in real time using an optical tachometer. The actual motor speed is provided as another input to the Opta PLC. The control algorithm compares desired to actual motor speed to update the PWM control signal. The PWM signal parameters (on time and off time) are varied, which adjusts the effective voltage supplied to the motor, to operate the motor at the desired speed.

Temp [oC]	V1	V2
-40	0.100	0.00
-35	0.150	0.30
-30	0.200	0.61
-25	0.250	0.91
-20	0.300	1.21
-15	0.350	1.52
-10	0.400	1.82
-5	0.450	2.12
0	0.500	2.42
5	0.550	2.73
10	0.600	3.03
15	0.650	3.33
20	0.700	3.64
25	0.750	3.94
30	0.800	4.24
35	0.850	4.55
40	0.900	4.85
45	0.950	5.15
50	1.000	5.45
55	1.050	5.76
60	1.100	6.06
65	1.150	6.36
70	1.200	6.67
75	1.250	6.97
80	1.300	7.27
85	1.350	7.58
90	1.400	7.88
95	1.450	8.18
100	1.500	8.48
105	1.550	8.79
110	1.600	9.09
115	1.650	9.39
120	1.700	9.70
125	1.750	10.00

K	6.060
Bias (B)	-0.606

TMP 36 temperature sensor
Temp in C = [(Vout in mV) - 500]/10

V2 = (V1 * K) + B

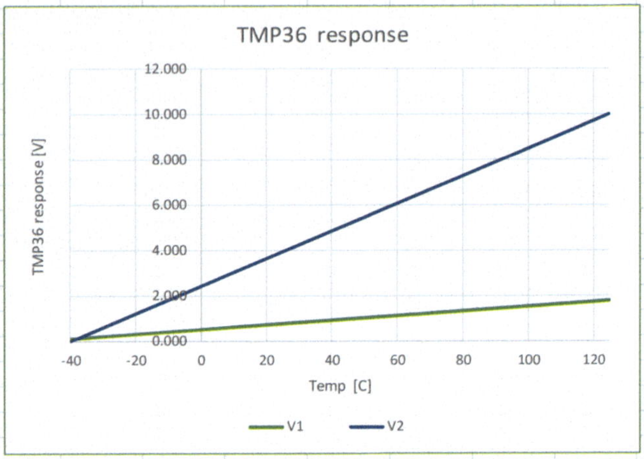

Fig. 4.35 TMP36 temperature sensor

This is provided as an illustrative example. It is not practical for a real world application due to the mechanical characteristics of the Opta's mechanical relay outputs. With that said, there are important concepts here to explore.[7]

The Opta PLC is not equipped with a dedicated hardware based pulse width modulation system. Instead, we use a technique called "bit banging" to emulate a PWM system using software. Bit banging may be employed in many applications when a software emulation replaces a dedicated hardware system (e.g. UART, SPI, etc.).

[7] We revisit this example in Chap. 6 using a dedicated PWM system.

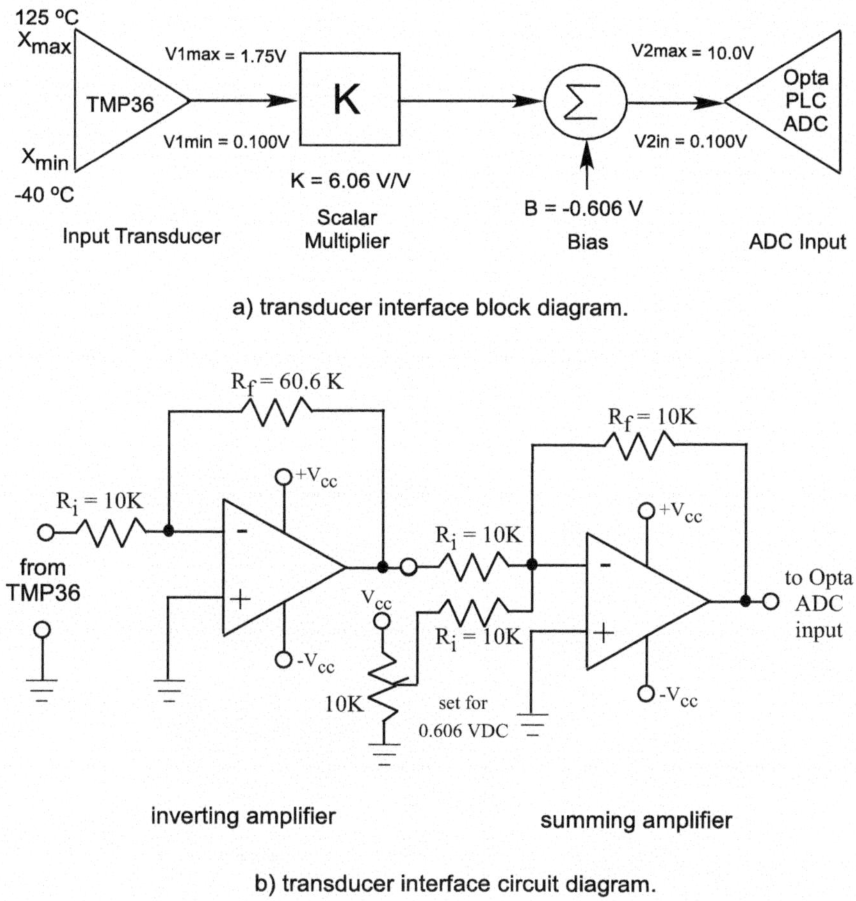

Fig. 4.36 TMP36 temperature sensor interface

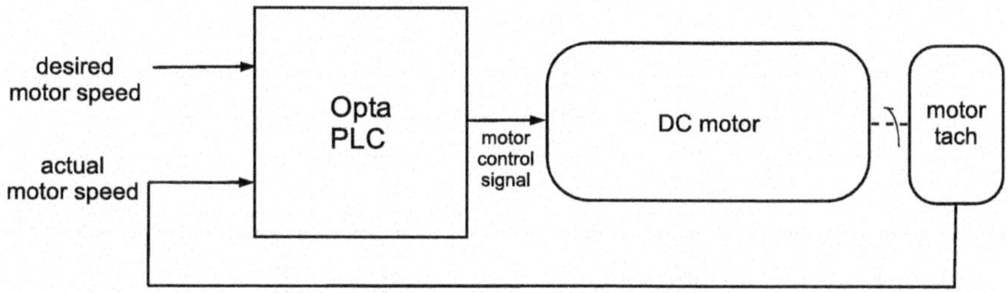

Fig. 4.37 Opta DC motor control. Images used courtesy of the Arduino team (CC BY–NC–SA) (www.arduino.cc)

4.16.1 Motor Control Hardware Configuration

The hardware configuration for the example is provided in Fig. 4.38. A DC–DC converter may be used to provided ±(VDC from 12 VDC for the op amp. This technique may be used in applications powered from a single polarity supply (e.g. greenhouse).

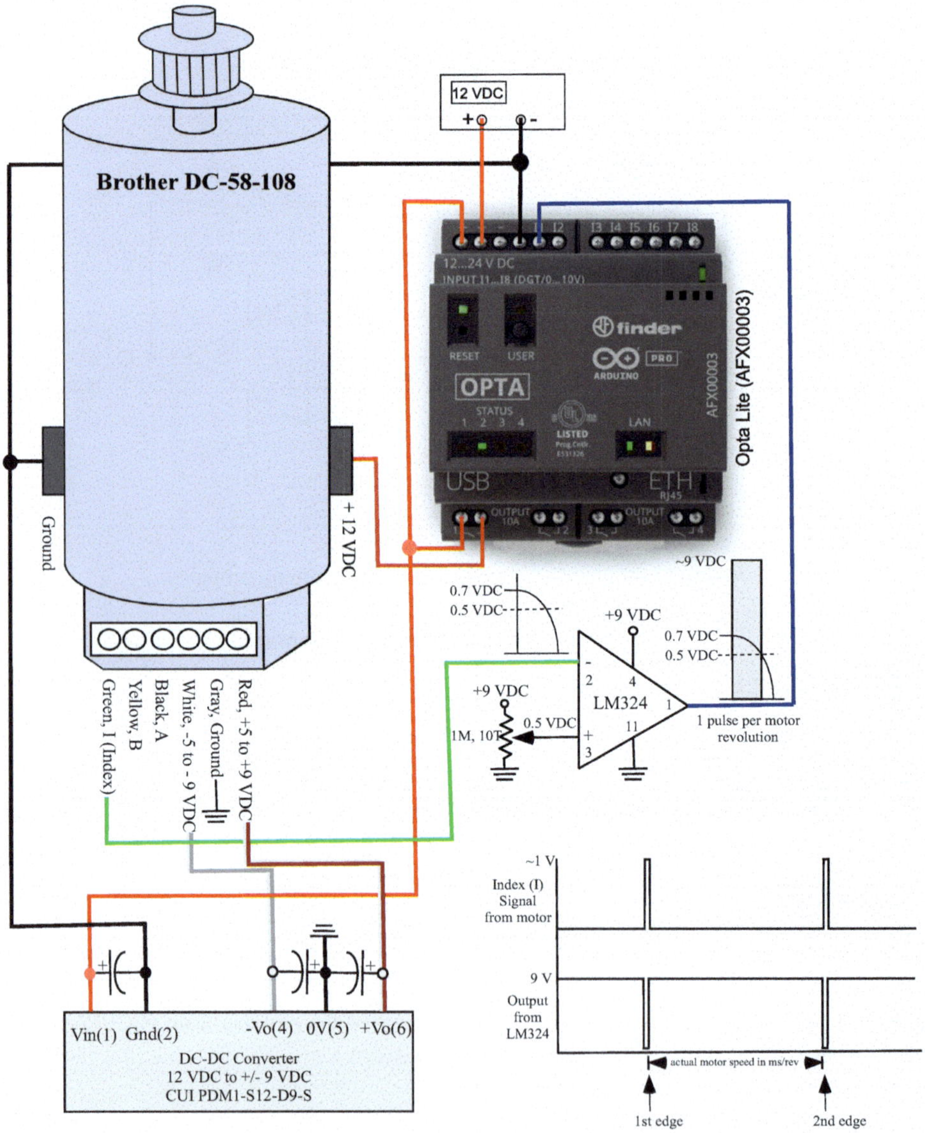

a) DC motor interface circuit. b) LM324 comparator signal conditioning.

Fig. 4.38 Opta DC motor control circuit

A Brother 12 VDC, 1,500 RPM DC motor equipped with an optical tachometer is used. The optical tachometer provides two sinusoidal outputs (A, B) and an index signal I. The signal I provides a 0.7 VDC index signal once per motor revolution. An LM324 op amp comparator circuit changes the 0.7 VDC signal to an approximate 9 VDC signal as shown in Fig. 4.39. The conditioned signal may be used as a direct measurement of actual motor speed by measuring the difference in time between the first and second edge.

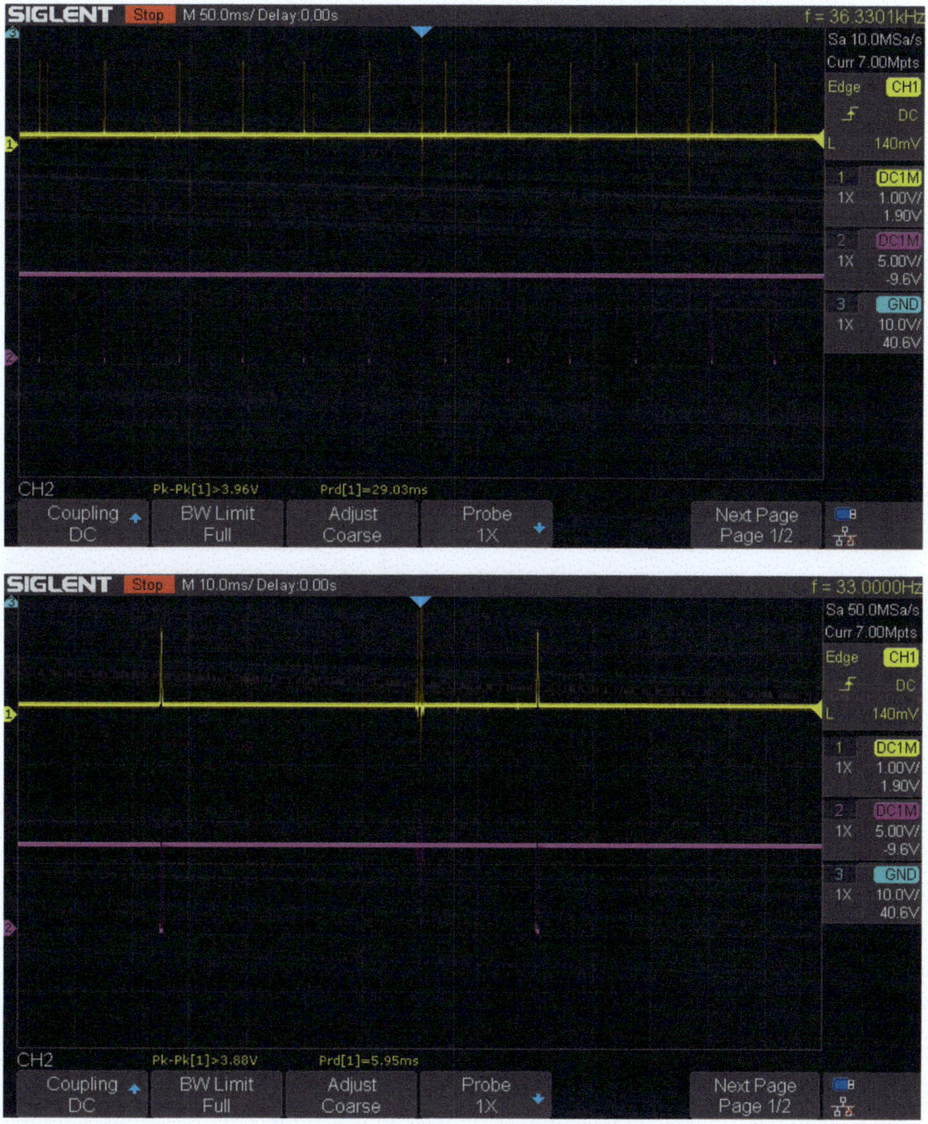

Fig. 4.39 Optical tachometer index pulse generator

4.16.2 Motor Control Software Configuration

The UML activity diagram for the motor control algorithm is provided in Fig. 4.40 and the accompanying Arduino IDE sketch is provided. The desired and actual motor speed are provided in units of ms/rev to be compatible with the optical tachometer output. When a difference between these two values is measured, motor speed is adjusted by varying the on and off time of the PWM signal.

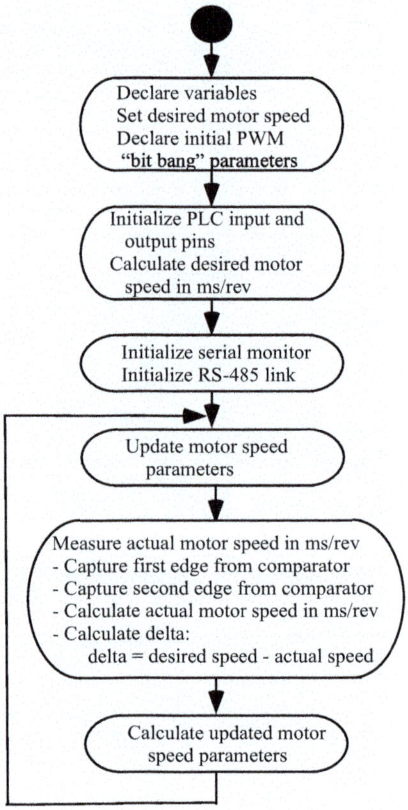

Fig. 4.40 Opta DC motor control UML

```
//*****************************************************************************
//Opta_motor_control
//- tracks and compensates for motor speed in ms/rev
//- 12 VDC, 1500 RPM motor
//- Desired motor speed: 1000 RPM
//- PWM duty cycle provides motor 8 VDC and ~1000 RPM
//- 1000 RPM = 16.7 RPS = 60 ms/rev
//
//S. Barrett, 01.08.24
//*****************************************************************************

int           motor_speed_pin = A0;           //monitors comparator output
double        desired_motor_rpm = 1000.0;     //motor speed in RPM
unsigned int  desired_motor_mspr;             //motor speed ms/rev
unsigned int  pulse_threshold = 1024;         //ADC level for high pulse
unsigned int  first_edge, second_edge;        //comparator signal edges
int           actual_motor_speed_mspr;        //measured motor speed [ms/rev]
int           delta = 0;                      //delta = desired - actual speed
int           delta_add = 0;                  //motor speed change
                                              //initial PWM parameters
int           motor_high_time = 100, motor_low_time = 40;
int           troubleshoot = 0;               //turns serial monitor on/off

void setup()
{
Serial.begin(9600);                           //serial monitor BAUD rate
analogReadResolution(12);                     //set 12 to 16 bits
pinMode(D0, OUTPUT);                          //Initialize Relays outputs
pinMode(LED_D0, OUTPUT);                      //Initialize Opta LEDs
pinMode(motor_speed_pin, INPUT);              //monitors comparator output
                                              //desired motor speed [ms/rev]
desired_motor_mspr = (unsigned int)((60.0/desired_motor_rpm)*1000.0);
                                              //update motor speed
digitalWrite(D0, HIGH);                       //Relay 1 on
digitalWrite(LED_D0, HIGH);                   //LED 1 on
delay(motor_high_time);                       //PWM: high baseline

digitalWrite(D0, LOW);                        //Relay 1 off
digitalWrite(LED_D0, LOW);                    //LED 1 off
delay(motor_low_time);                        //PWM: low baseline
}

void loop()
{
if(troubleshoot)
  {
  Serial.print("Desired motor speed [ms/rev]:");
  Serial.println(desired_motor_mspr);
  }
                                              //Update motor speed
digitalWrite(D0, HIGH);                       //Relay 1 on
digitalWrite(LED_D0, HIGH);                   //LED 1 on
motor_high_time = motor_high_time + delta_add; //PWM: high update
if(motor_high_time < 0)  motor_high_time = 0;
delay(motor_high_time);
```

```
if(troubleshoot)
  {
  Serial.print("High time:");
  Serial.println(motor_high_time);
  }

                                               //process pin while high
while(analogRead(motor_speed_pin) >= pulse_threshold)
  {
  ;
  }
                                               //Measure motor speed [ms/rev]
                                               //Get time hack first edge
                                               //process pin while low
while(analogRead(motor_speed_pin) < pulse_threshold)
  {
  ;
  }
first_edge = millis();                         //capture first edge
                                               //process pin while high
while(analogRead(motor_speed_pin) >= pulse_threshold)
  {
  ;
  }
second_edge = millis();                        //capture second edge

if(troubleshoot)
    {
    Serial.print("Second edge: ");
    Serial.println(second_edge);
    }
                                              //measured motor speed [ms/rev]
actual_motor_speed_mspr = second_edge - first_edge;

if(troubleshoot)
  {
  Serial.print("Actual motor speed: ");
  Serial.println(actual_motor_speed_mspr);
  }

//Calculate delta
//+ delta: motor too slow, speed up
//- delta: motor too fast, slow down
delta = desired_motor_mspr - actual_motor_speed_mspr;

if(troubleshoot)
  {
  Serial.print("Delta: ");
  Serial.println(delta);
  }

if(delta > 0)                                  //motor slow, speed up
  delta_add = - delta/2;
else if (delta < 0)                            //motor fast, slow down
  delta_add = + delta/2;
else                                           //motor at desired speed
```

```
  delta_add = 0;

digitalWrite(D0, LOW);                              //PWM: low time update
digitalWrite(LED_D0, LOW);
motor_low_time = motor_low_time - delta_add;
if(motor_low_time < 0)  motor_low_time = 0;
delay(motor_low_time);

if(troubleshoot)
  {
  Serial.print("Low time:");
  Serial.println(motor_low_time);
  }

if(troubleshoot)
  {
  Serial.print("Delta add: ");
  Serial.println(delta_add);
  Serial.println();
  }

}

//*********************************************************************
```

4.17 Summary

In this chapter we explored how an Opta PLC may be used in industrial, Internet of Things (IoT), and hybrid applications. We explored how to connect input sensors and output actuators to an Opta PLC. We began with a review of the Opta input and output characteristics. We then explored a wide variety of digital and analog input sensors and output actuators. We also employed an operational amplifier–based transducer interface design process to interface input sensors to the Opta PLC.

4.18 Problems

1. What will happen if a controller is used outside of its prescribed operating envelope?
2. Discuss the difference between the terms "sink" and "source" as related to current loading of a microcontroller.
3. What is switch bounce? Describe techniques to minimize switch bounce.
4. What is the difference between an incremental encoder and an absolute encoder? Describe applications for each type.
5. Describe an application for a flex sensor. Provide a supporting circuit and software with all components specified.

6. Describe an application for a fluid level sensor. Provide a supporting circuit with all components specified.
7. Why is a transimpedance amplifier typically used with a photodiode circuit?
8. What are the advantages and disadvantages of using the Arduino Opta PLC relay outputs in a given application?
9. Construct a table of different switch types. Provide an application for each type.
10. What is the purpose of a two channel incremental quadrature encoder?
11. What is ultrasound? How might an ultrasound sensor be used in an industrial environment?
12. What is the advantage of the TMP36 temperature sensor over other sensors that measure negative temperatures?
13. Describe an industrial application for a tilt sensor.
14. Describe an application for an optical isolator circuit. Provide a supporting circuit with all components specified.
15. Describe an application for an environmental sensing circuit. Provide a supporting circuit with all components specified.
16. What are the ideal operational amplifier characteristics? What prevents an op amp from performing in an ideal manner?
17. In your own words describe each of the op amp nonideal parameters.
18. Derive each of the output equations for the classic operational amplifier configurations.
19. Draw a UML activity diagram for the LM34 ladder logic program.
20. For the LM34 ladder logic program set the fan on and off thresholds with a potentiometer.
21. Write and test a ladder logic program for the linear actuator circuit discussed in the chapter.

References

Advanced Photonix Incorporated (API), *CdS Photoconductive Photocells PDV–P8001,* March 2006.

Arduino Opta Collective Datasheet, Product Reference Manual, SKU: AFX00001– AFX00002– AFX00003, www.arduino.cc, June 2023.

S. BharadwajRedd, *PLC Programming for Blinking Indicator Lights,* www.learn.automationcommunity.com

D. Egironi, *Presenting MQ sensors: low–cost gas and pollution detectors*, Open Electronics Source Electronic Projects, February 2018.

Industrial Fiber Optics, i-fiberoptics.com.

LM34 Precision Fahrenheit Temperature Sensors, Texas Instruments SNIS161D, March 2000–Revised January 2016, www.ti.com.

LM35 Precision Centigrade Temperature Sensors, Texas Instruments SNIS159H, August 1999–Revised January 2017, www.ti.com.

Low Voltage Temperature Sensors, TMP35/36/37, Analog Devices, Rev. H, 2015, www.analog.com.

Franklin Miller, Jr. *College Physics,* fourth edition, Harcourt Brace Jovanovich, Inc. 1977.

Milone Technologies, *0–5 VDC Linear Resistance to Voltage Module PN–05V00199 Rev 2*, www.milonetech.com.

Milone Technologies, *eTape Continuous Fluid Level Sensor PN–12110215TC–X*, www.milonetech.com.

MQ—2 Semiconductor Sensor for Combustible Gas,, www.mysensors.org

MQ—3 Gas Sensor, Hanwei Electronics Co., ltd., www.hwsensor.com.

MQ—4 Gas Sensor, Hanwei Electronics Co., ltd., www.hwsensor.com.

MQ—6 Flammable Gas Sensor, Zhengzhou Winsen Electronics Technology Co., Ltd, www.winsensor.com

MQ—7 Gas Sensor, Hanwei Electronics Co., ltd., www.hwsensor.com.

MQ—8 Flammable Gas Sensor, Zhengzhou Winsen Electronics Technology Co., Ltd, www.winsensor.com.

MQ—9 Semiconductor Sensor for CO/Combustible Gas, Hanwei Electronics Co., ltd., www.hwsensor.com.

MQ–131 Gas Sensor, www.mysensors.org.

MQ–135 Gas Sensor, www.mysensors.org.

MQ–183 Gas Sensor, Hanwei Electronics Co., ltd., www.hwsensor.com.

MQ–214 Gas Sensor, Hanwei Electronics Co., ltd., www.hwsensor.com.

Sick/Stegmann Incorporated, Dayton, OH, (www.stegmann.com).

SparkFun Electronics, 6175 Longbow Drive, Suite 200, Boulder, CO 80301 (www.sparkfun.com)

uA7800 series positive–voltage regulators, SLVS056J, Texas Instruments, 2003.

L.M. Faulkenberry, *An Introduction to Operational Amplifiers,* John Wiley & Sons, 1977.

W. Jung, *Op Amp History,* Analog Devices www.analog.com.

J.R. Ragazzini, R.H. Randall, and F.A. Russell, *Analysis of Problems in Dynamics by Electronic Circuits,* Proceedings of the I.R.E., 444–452, May 1947.

A.S. Sedra and K.C. Smith, *Microelectronics,* Oxford University Press, 2004.

D.F. Stout and M. Kaufman, *Handbook of Operational Amplifier Circuit Design,* McGraw–Hill Book Company, 1976.

Application: IoT Greenhouse 5

Objectives: After reading this chapter, the reader should be able to do the following:

- Apply instrumentation, IoT, and Opta PLC concepts to design a control system;
- Describe tools used to systematically design hardware and software tools for a control project; and
- Implement the control system with both the Arduino IDE and PLC IDE.

5.1 Objective

The objective of this chapter is to demonstrate in action the concepts discussed in this book. Simply put, our goal is to provide the theory, design, and construction of a passively heated greenhouse. We equip the greenhouse with instrumentation to monitor and control key parameters. Using IoT concepts key parameters will be monitored and controlled remotely. Via a greenhouse controller example we demonstrate how to interface different sensors and actuators to an Arduino Opta PLC. These examples may be used and adapted for many other non–greenhouse systems.[1]

We begin the chapter with the theory of greenhouse design. The reader is assumed to have no background in this area. This section was compiled from a number of excellent sources listed at the end of the chapter. We then present the design and construction of a passively heated greenhouse. In our example an existing eight by eight foot garden shed is converted into a greenhouse. This is a do it yourself (DIY) project.

[1] This chapter was adapted for the Arduino Opta PLC with permission from "Internet of Things," S. Barrett, Springer, 2021.

S. F. Barrett, *Arduino VII*, Synthesis Lectures on Digital Circuits & Systems,
https://doi.org/10.1007/978-3-031-68609-2_5

Prior to this project, the author had no background in greenhouse design and construction. However, it was a project considered for some time. The chapter then provides details on an Arduino Opta based greenhouse control system. Finally, key greenhouse parameters are made available for monitoring and control using Arduino–based IoT concepts. Prior to continuing, the reader is encouraged to review microcontroller–based system design concepts provided in Appendix B.

5.2 Greenhouse Theory

There is considerable information available to guide the design of a passive greenhouse. A passive greenhouse uses solar energy to either extend the growing season for plants or to grow plants year round. Equally important is to provide for greenhouse cooling and ventilation during hot summer months. The information provided here is a compilation of the excellent sources listed at the end of this chapter. A thorough review of these sources is recommended.

Passive heating uses the energy of the sun to heat the interior of the greenhouse during the day. The heat energy is stored using a large thermal mass such as barrels filled with water as shown in Fig. 5.1. When the temperature drops at night, the energy stored in the water is released into the greenhouse to mitigate internal temperature fluctuations.

The British Thermal Unit (BTU) is the energy required to raise one pound of water a degree Fahrenheit. So during the day, the water barrels absorb energy. At night a drop of one degree per pound of water would release one BTU of heat energy.

To efficiently capture the solar energy, greenhouse windows should face south. Windows may be included on the east and west sides as well. Typically, the north facing wall is not equipped with windows but is thoroughly insulated.

The south facing windows ideally should be inclined at an angle. A general rule of thumb is to take your location's latitude and add ten degrees. I live in Laramie, Wyoming ($41.3114^\circ\ N,\ 105.5911^\circ\ W$); therefore, ideally the south facing windows should be inclined at 51°.

To determine the amount of water required for passive energy storage several rules of thumb are used:

- To extend the growing season, 2.5 gallons of water needs to be stored for every square foot of glazing material (windows).
- To grow plants year round, five gallons of water needs to be stored for every square foot of glazing material.
- It is worth noting that a gallon of water weighs 8.3 pounds.

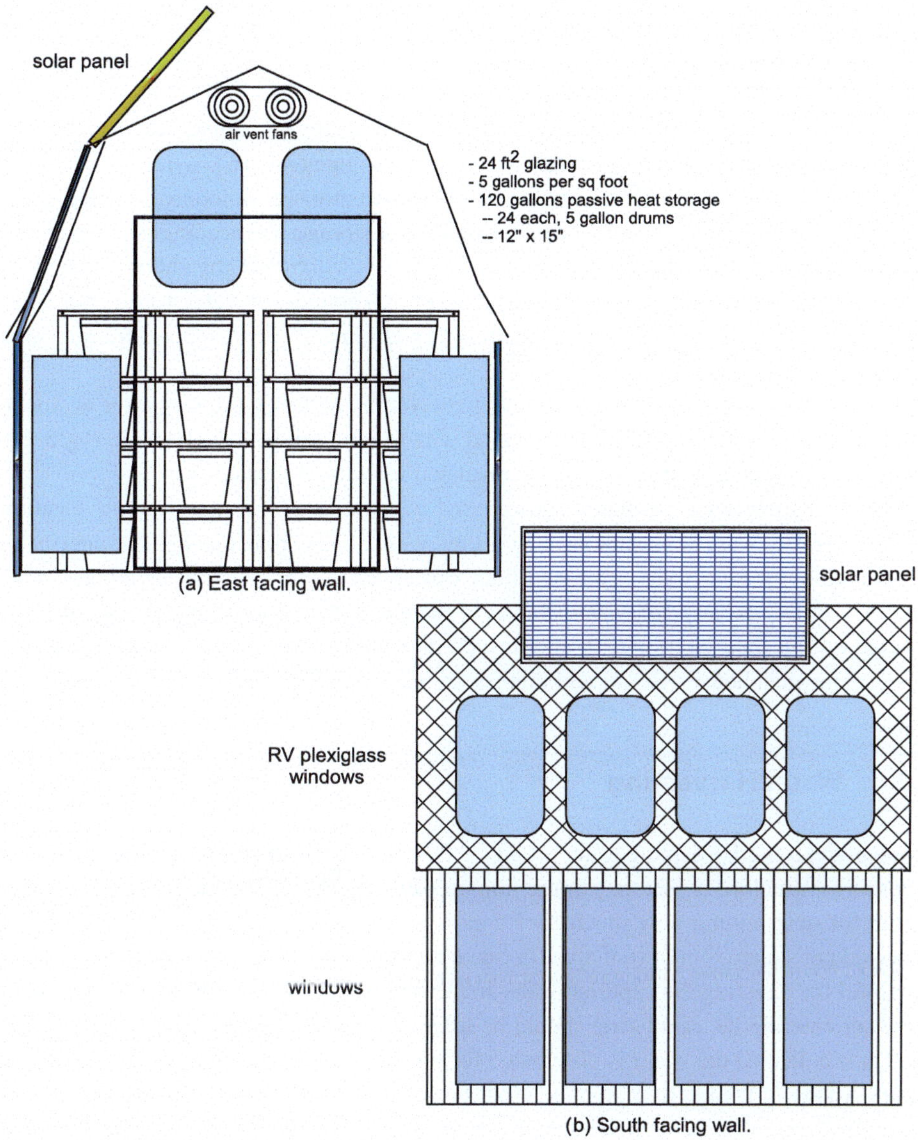

Fig. 5.1 Greenhouse concepts

To maximize the collection of solar energy, it was decided to place windows on the south, east, and west walls of the greenhouse. In this specific example, an eight foot by eight foot by eight foot existing garden shed was converted to a greenhouse.

A combination of mobile home glass windows and plexiglass bubble (RV skylight) windows were installed. The total glazing area was 36.2 square feet. To provide for year round growing, 180 gallons of water is required for passive storage. A total of 24 five gallon, black plastic buckets were used. The balance of water storage was accomplished using a 55 gallon rain barrel. The buckets were placed on shelves within the greenhouse. Realize the greenhouse floor will be supporting 180 gallons (1,500 pounds) of water. Ensure the floor is adequately supported to do this. I included some additional 2" × 4" support framing under the floor for this purpose.

The shed has traditional 2" × 4" framing covered by half inch thick exterior wallboard. The roof of the shed was peaked and covered with roofing shingles as shown in Fig. 5.2 top left. The north facing wall of the shed is insulated with R–13 insulation.

The shed's interior is insulated and covered with wallboard in non–window locations. The interior roof was also insulated and covered with 1" × 4" pine planks. The interior was painted white for reflectivity. The shed floor was covered with a quarter inch black rubber mat. A 12 VDC exhaust fan was installed in the east facing eave. Finally, windows were covered on the shed interior with eighth inch plexiglass to provide an additional thermal barrier. The greenhouse exterior and interior are shown in Fig. 5.2.[2]

5.3 Water Harvesting

A rain barrel is used to capture water for greenhouse use. Typically the roof of the greenhouse or a nearby structure is used to capture rain and direct it to a containment barrel. A rule of thumb for determining how much roof area is needed is a half–gallon of water may be collected per square foot of roof area for one inch of rain fall. The gutter system on the roof is helpful for directing the captured water to the rain barrel.

Water entering the rain barrel should be filtered to remove leaves, twigs, roof material, etc. Once collected the water is distributed for greenhouse use by employing a small water pump. A gravity feed system may be used; however, the rain barrel may need to be set at a prohibitive height to obtain suitable water pressure (Watson).[3] For this project we use a commercially available 50 gallon, plastic rain barrel for water catchment.

[2] Springer is a global publisher. Due to the wide range of applicable building codes and standards and permitting requirements, please check with local requirements concerning permitting and installation requirements.

[3] Check local rules and codes to determine if water harvesting is permitted in your location.

Fig. 5.2 Greenhouse project

5.4 Greenhouse Control System Requirements

The Greenhouse Control System (GCS) has the following requirements:

- Self–contained solar power with a 12 VDC battery backup,
- Compartmentalized for systematic design and expansion, and

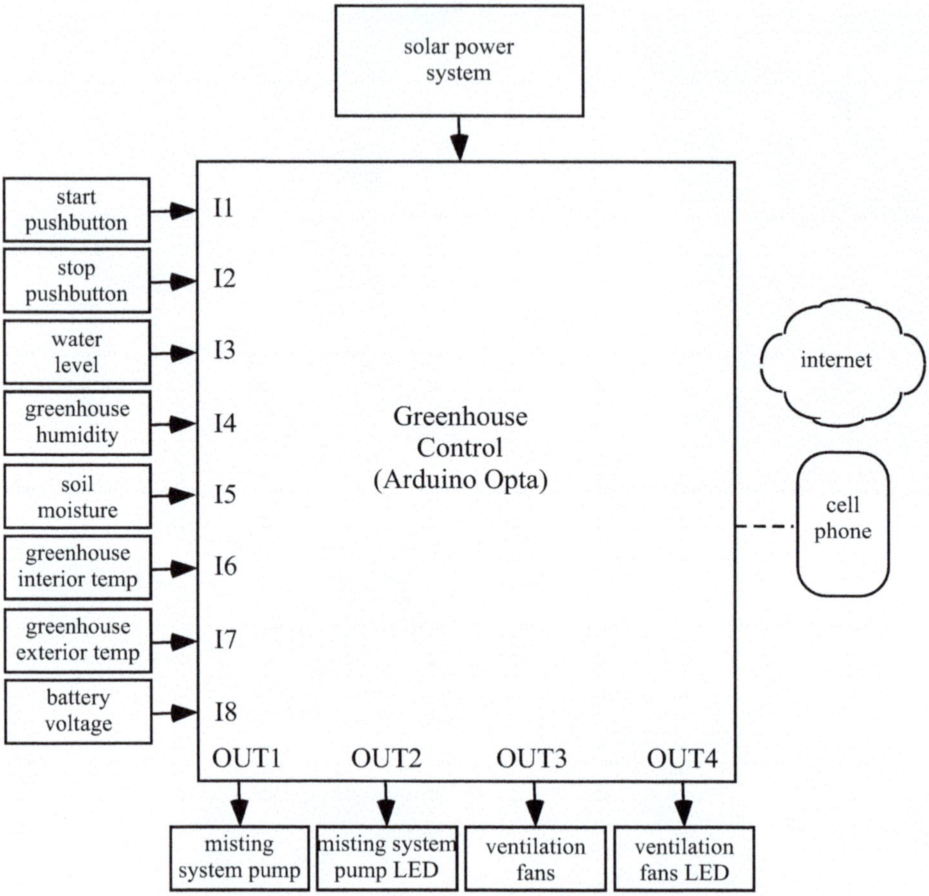

Fig. 5.3 Greenhouse control system

- Arduino Opta PLC based technology for greenhouse instrumentation and control.

The GCS block diagram is shown in Fig. 5.3.
In keeping with a compartmentalized design, the GCS has three subsystems:

- Solar power system;
- Greenhouse Control System; and
- Arduino Opta PLC based IoT interface.

Each subsystem is discussed in turn.

5.5 Solar Power System

The Solar Power Systems was discussed in "Arduino I: Getting Started." An excerpt is included here with permission for completeness.

The solar power system consists of a solar panel, a solar power manager, a rechargeable battery, and fuses for circuit protection. In this project we use the DFRobot DFR0580 Solar Power Manager for a 12 VDC lead–acid battery. With an 18 VDC, 100 W solar panel and a 12 VDC lead–acid battery; the DFR0580 can provide regulated output voltages of 5 VDC at 5 amps and 12 VDC at 8 amps. This is suitable for the GCS project (www.DFRobot.com). A diagram of the solar power system is shown in Fig. 5.4.

A distribution panel was designed for the power system and is shown in Fig. 5.5. The various components are connected together as shown in the figure with an automotive style fuse block and automotive style blade fuses. The panel is housed in a QILIPSU plastic, hinged 16.1" × 12.2" × 7.1" enclosure.

5.6 Greenhouse Control System

As shown in Fig. 5.3, the following requirements have been set for the Greenhouse Control portion of the GCS system:

- Monitor the water level in the rain barrel (analog input I3);
- Monitor humidity level within the greenhouse (analog input I4);
- Monitor plant soil moisture (analog input I5);
- Monitor indoor greenhouse temperature (analog input I6);
- Monitor outdoor greenhouse temperature (analog input I7);
- Monitor battery voltage level (analog input I8);
- Activate vent fan, vent van LED, and open vents when the internal greenhouse temperature is above a desired value (relay output 1);

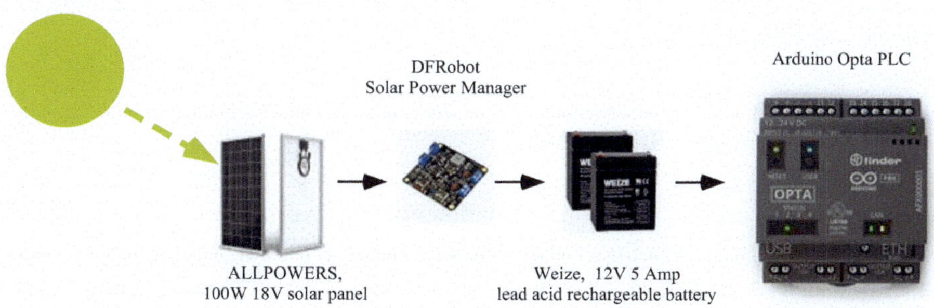

Fig. 5.4 Solar power system. Images courtesy of AllPowers, DFRobot, Weize, and Arduino

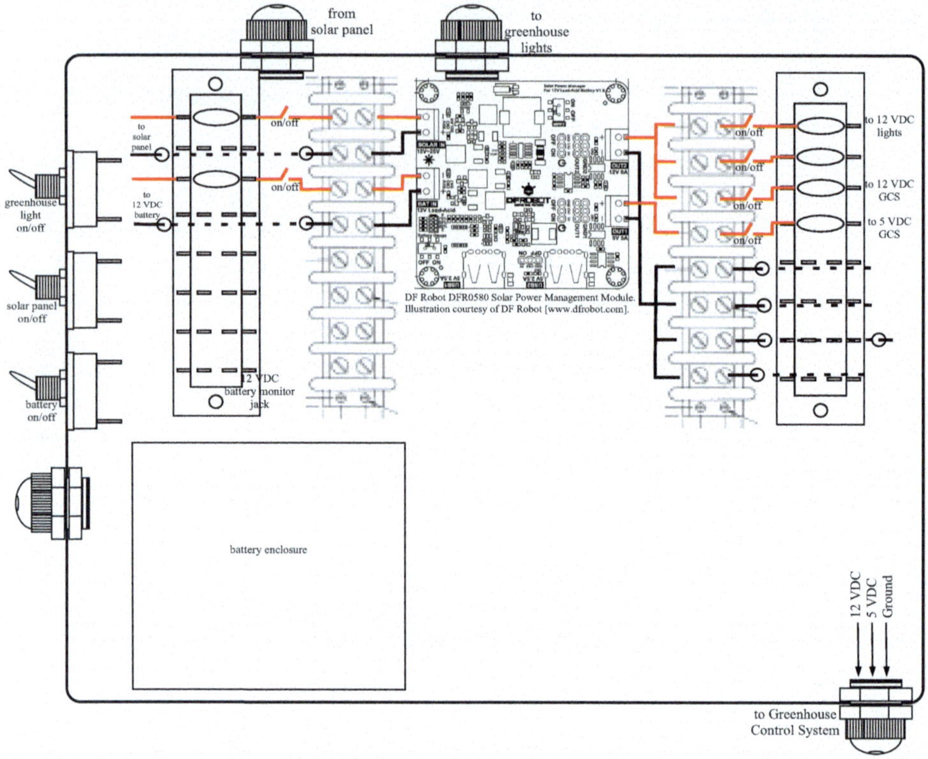

Fig. 5.5 Solar power distribution panel. Arduino illustrations used with permission of the Arduino team (CC BY–NC–SA) [www.arduino.cc]

- Activate misting system, misting system LED when the internal humidity level falls below the desired value and/or plant soil moisture is low (relay output 2); and
- Provide two status LEDs S1 and S2 to report program status (relay outputs 3 and 4).

To meet these requirements the instrumentation system shown in Fig. 5.6 is used. Each component is discussed in turn.

5.6.1 Milone E–Tape Fluid Sensor

Milone Technologies manufacture a line of continuous fluid level sensors. The sensor resembles a ruler and provides a near linear response. The sensor reports a change in resistance to indicate the distance from sensor top to the fluid surface. To convert the resistance change to

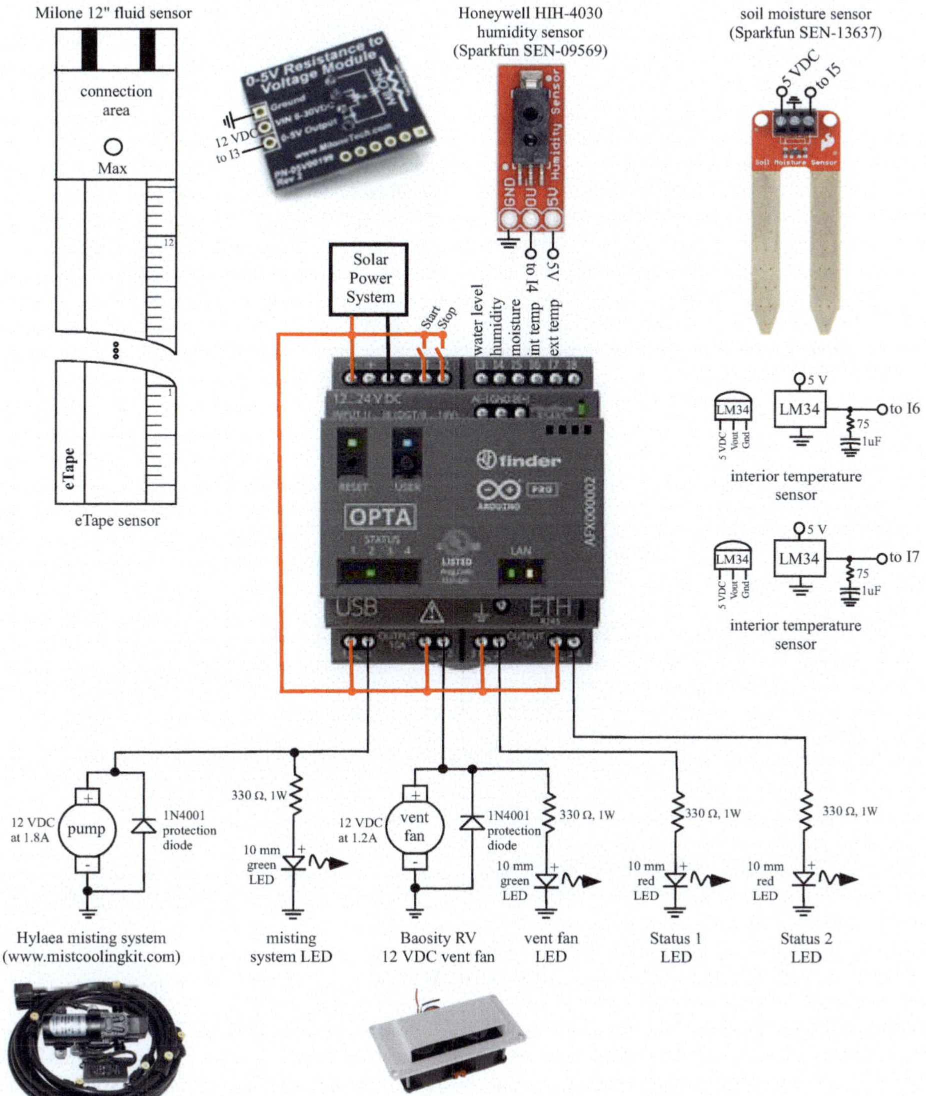

Fig. 5.6 Greenhouse control system. Arduino illustrations used with permission of the Arduino team (CC BY–NC–SA) [www.arduino.cc]

a voltage change, the Milone 0–5 VDC Resistance to Voltage Module is used. The module shown in Fig. 5.7a(left) is powered from 12 VDC. The output from the module ranges up to 5 VDC (www.milonetech.com). A sample sketch to collect data from this sensor is provided in the Application section of the chapter.

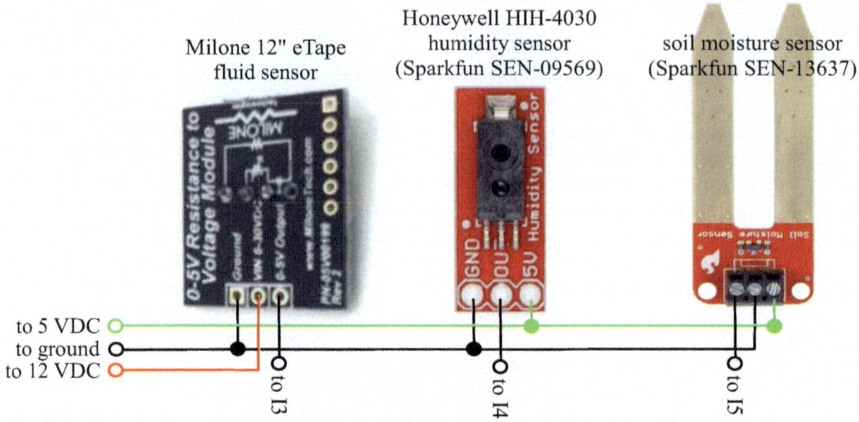

(a) GCS sensors.

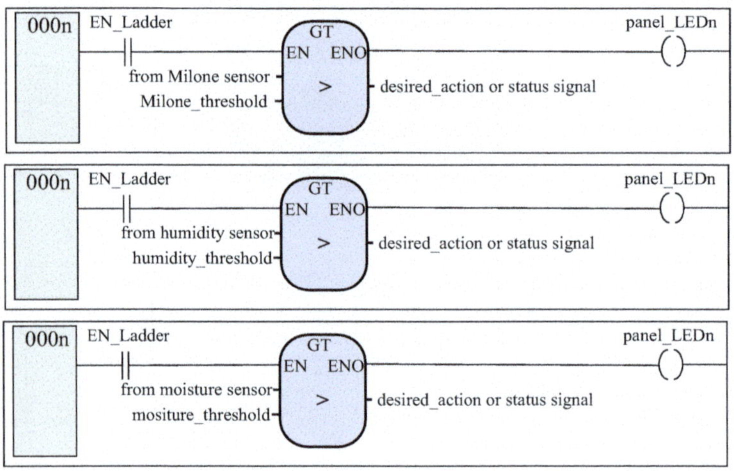

(b) ladder logic examples.

Fig. 5.7 (left)Milone Technologies eTape liquid level sensor. Image courtesy of Milone Technology (www.milonetech.com). (center) Honeywell HIH–4030 sensor. Image courtesy of Sparkfun (CY BY 2.0) (www.sparkfun.com). (right) Sparkfun soil moisture sensor. Image courtesy of Sparkfun (CY BY 2.0) (www.sparkfun.com)

5.6.2 Humidity Sensor

A Honeywell HIH–4030 sensor is used to measure greenhouse humidity. The sensor provides an output voltage that may be mapped to a corresponding relative humidity (RH) value. The RH value provides a measurement of the mount of water vapor in the air. The RH is expressed as a value from 0 to 100 percent RH. The interface circuit for the RH sensor is shown in Fig. 5.7a(middle).

The sensor provides an output voltage to indicate RH. The voltage is processed and corrected for temperature using the following equations provided by the manufacturer (Honeywell).

$$V_{out} = (V_{supply}) * (0.0062 * sensorRH) + 0.16$$

The sensor RH value is corrected for temperature:

$$TrueRH = sensorRH/(1.0546 - 0.00216T)$$

with T expressed in degrees Centigrade. A sample sketch to collect data from this sensor is provided in the Application section of the chapter.

5.6.3 Soil Moisture Sensor

A Sparkfun soil moisture sensor (SEN–13637) is used to monitor plant soil moisture content. The sensor is powered by a 5 VDC source. The interface circuit is shown in Fig. 5.7a(right). Sensor sample ladder logic code is provided Fig. 5.7b. A sample sketch to collect data from this sensor is provided in the Application section of the chapter.

5.6.4 LM34 Interior Greenhouse Temperature Sensor

To monitor the interior and exterior greenhouse temperature a pair of LM34 Precision Fahrenheit Temperature sensors are used. The LM34 provides 10 mV of output per degree Fahrenheit. As configured, the LM34s report Fahrenheit temperatures down to zero degrees. A design to provide temperature sensing below zero degrees is provided as an end of chapter exercise. The LM34 measuring external temperature should be housed in a weather protective enclosure as shown in Fig. 5.8. Sample ladder logic code is provided for the sensors in Fig. 5.8c.

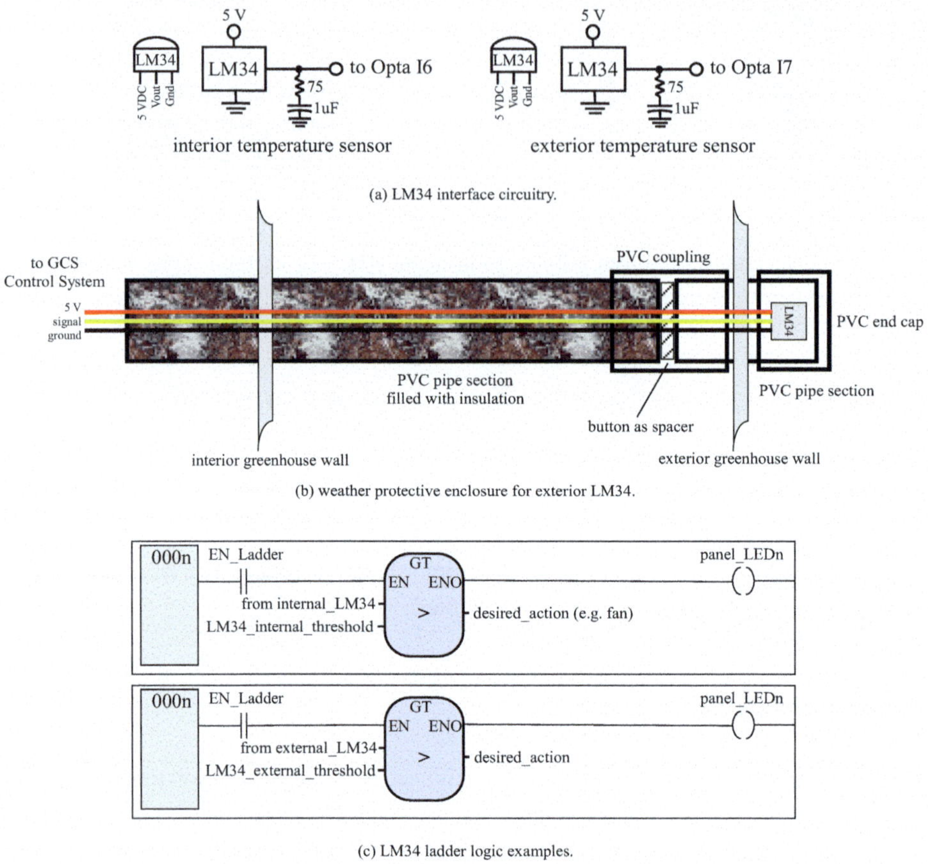

Fig. 5.8 Interior and exterior greenhouse monitor

5.6.5 Misting System and LED

The greenhouse misting system is shown in Fig. 5.9. A Hylaea misting system consisting of a 12 VDC fluid pump and misting delivery hardware is provided in the kit (www.mistcoolingkit.com).

A 1N4001 diode serves as protection for the inductive load. The 12 VDC source is supplied by the DF Robot DFR0580 Solar Power Management Module OUT 2 rated at 12V, 8A. An LED indicator circuit is also provided to indicate when the pump is running.

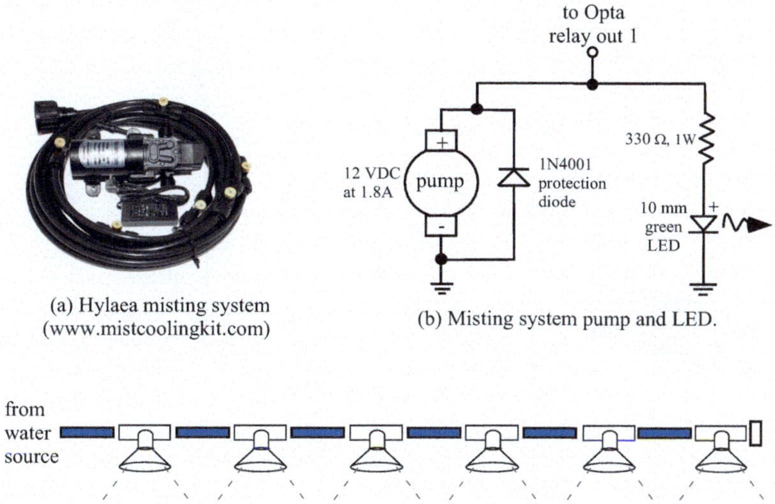

(a) Hylaea misting system (www.mistcoolingkit.com)

(b) Misting system pump and LED.

(c) misting delivery system

Fig. 5.9 Hylaea misting system

5.6.6 Vent Fan and LED

The greenhouse vent fan system is shown in Fig. 5.10. A Baosity recreational vehicle 12 VDC vent fan is mounted in the east eave of the greenhouse roof. It provides for safe venting of the greenhouse interior when it becomes too hot.

A 1N4001 diode serves as protection for the inductive load. The 12 VDC source is supplied by the DF Robot DFR0580 Solar Power Management Module OUT 2 rated at 12 V, 8 A. An LED indicator circuit is also provided to indicate when the vent is running.

5.6.7 GCS System Code

The UML activity diagram for the GCS system code is provided in Fig. 5.11. An Arduino IDE sketch follows. It provides a template for basic control system operation. It may be customized for your local climate conditions.

```
//*****************************************************************
//Opta_GH_control
//
//Monitors:
//I1: Start - Green PB (digital)
//I2: Stop - Red PB (digital)
```

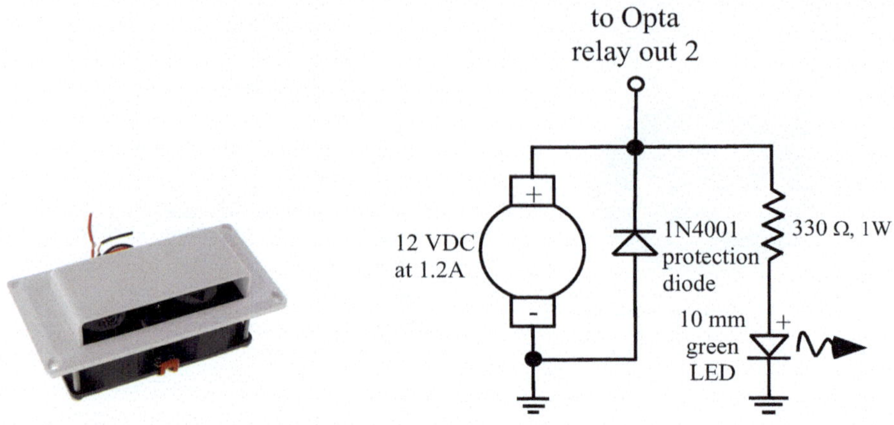

(a) Baosity RV 12 VDC vent fan

(b) 12 VDC vent fan and LED.

Fig. 5.10 Greenhouse vent system

```
//I3: Water level (analog) - Milone e-Tape sensor
//I4: Humidity (analog) - Honeywell HIH sensor
//I5: Soil moisture (analog) - Sparkfun moisture sesnor
//I6: Internal temperature (analog) - LM34
//I7: External temperature (analog) - LM34
//I8: Battery voltage level (analog) - 1/2V from 12V lead acid batt
//
//Controls:
//Out 1: Water pump for misting system
//Out 2: Vent fan for excessive internal temperature
//
//Status:
//Out 1: Pump LED
//Out 2: Vent fan LED
//Out 3: Status 1 LED
//Out 4: Status 2 LED
//
//Notes:
//- Opta Wi-Fi required for BlueTooth BLE link
//- Thresholds based on 12 bit ADC, Vmax = 10 VDC
//
//Last revised:  02.02.24  S. Barrett
//****************************************************************

unsigned int troubleshoot = 1;       //1: serial monitor status prints
                                     //set status thresholds
unsigned int keep_going = 0;         //loop variable for start/stop
```

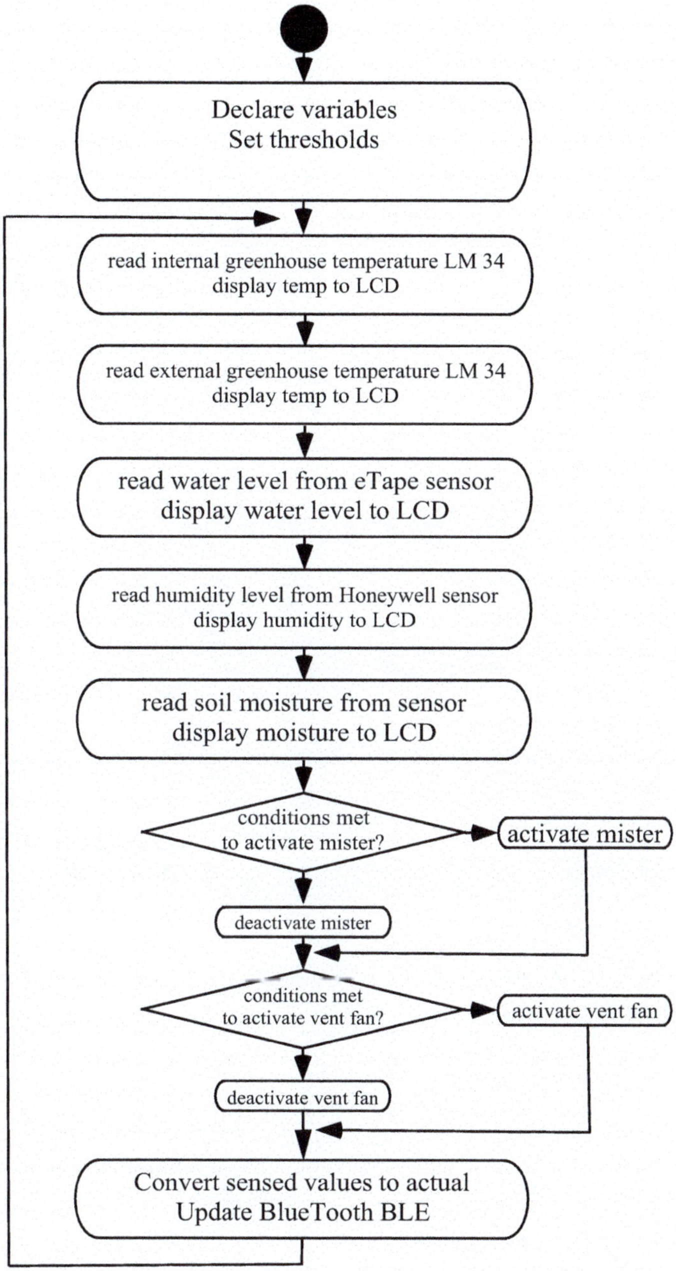

Fig. 5.11 Greenhouse UML activity diagram

```
unsigned int water_level_th = 1024; //water level threshold
unsigned int humidity_th = 1024;    //humidity level threshold
unsigned int moisture_th = 1024;    //mositure level threshold
unsigned int int_temp_th = 370;     //GH int temp threshold
unsigned int ext_temp_th = 370;     //GH ext temp threshold
unsigned int batt_lvl_th = 2458;    //batt level threshold
                                    //status values
unsigned int water_level;           //water level value
unsigned int humidity;              //humidity level value
unsigned int moisture;              //mositure level value
unsigned int int_temp;              //GH int temp value
unsigned int ext_temp;              //GH ext temp value
unsigned int batt_lvl;              //batt level value
unsigned int loop_cnt =0;           //loop counter

void setup()
{
if(troubleshoot)Serial.begin(9600); //configure Serial Monitor

analogReadResolution(12);           //ADC set 12 to 16 bits
pinMode(PIN_A0, INPUT);             //Init Start input I1
pinMode(PIN_A1, INPUT);             //Init Stop input I2

pinMode(LED_D0, OUTPUT);            //Initialize status LEDs 1 to 4
pinMode(LED_D1, OUTPUT);
pinMode(LED_D2, OUTPUT);
pinMode(LED_D3, OUTPUT);

pinMode(D0, OUTPUT);                //Initialize relays outputs
pinMode(D1, OUTPUT);
pinMode(D2, OUTPUT);
pinMode(D3, OUTPUT);
}

void loop()
{
if(troubleshoot) Serial.print("loop begin: ");
if(troubleshoot) Serial.println(loop_cnt);
if((digitalRead(PIN_A0))||(keep_going))//start or keeping going
  {
  keep_going = 1;                   //seal start PB
  if(troubleshoot) Serial.println("keep going");

  //Update status
  water_level = analogRead(A2);     //I3
  humidity    = analogRead(A3);     //I4
  moisture    = analogRead(A4);     //I5
  int_temp    = analogRead(A5);     //I6
  ext_temp    = analogRead(A6);     //I7
```

```
batt_lvl      = analogRead(A7);      //I8

if(troubleshoot)
  {
  if(troubleshoot) Serial.print("water level: ");
  if(troubleshoot) Serial.println(water_level);
  if(troubleshoot) Serial.print("humidity: ");
  if(troubleshoot) Serial.println(humidity);
  if(troubleshoot) Serial.print("moisture: ");
  if(troubleshoot) Serial.println(moisture);
  if(troubleshoot) Serial.print("int temp: ");
  if(troubleshoot) Serial.println(int_temp);
  if(troubleshoot) Serial.print("ext temp: ");
  if(troubleshoot) Serial.println(ext_temp);
  if(troubleshoot) Serial.print("batt lvl: ");
  if(troubleshoot) Serial.println(batt_lvl);
  }

//Update controls and status LEDs

//vent fan
if(int_temp >= int_temp_th)
  {
  digitalWrite(D1, HIGH);        //vent van on
  if(troubleshoot) Serial.println("vent fan on");
  }
else
  {
  digitalWrite(D1, LOW);         //vent van off
  if(troubleshoot) Serial.println("vent fan off");
  }

//soil moisture
if(moisture <= moisture_th)
  {
  digitalWrite(D0, HIGH);        //mist pump on
  if(troubleshoot) Serial.println("mist pump on");
  }
else
  {
  digitalWrite(D0, LOW);         //mist pump off
  if(troubleshoot) Serial.println("mist pump off");
  }

  //battery level
  if(batt_lvl <= batt_lvl_th)
  {
  digitalWrite(D2, HIGH);        //Status LED 1 on
  if(troubleshoot) Serial.println("battery low");
```

```
      }
    else
      {
      digitalWrite(D2, LOW);          //Status LED 1 off
      if(troubleshoot) Serial.println("battery ok");
      }

    //Convert sensor readings to actual values

    //Update BLE status report

    }
  else
    {
    if(troubleshoot) Serial.println("keep going - no");
    keep_going = 0;                      //stop
    //flash Opta panel LEDs
    digitalWrite(LED_D0, HIGH);
    delay(100);
    digitalWrite(LED_D0, LOW);
    delay(100);

    digitalWrite(LED_D1, HIGH);
    delay(100);
    digitalWrite(LED_D1, LOW);
    delay(100);

    digitalWrite(LED_D2, HIGH);
    delay(100);
    digitalWrite(LED_D2, LOW);
    delay(100);

    digitalWrite(LED_D3, HIGH);
    delay(100);
    digitalWrite(LED_D3, LOW);
    delay(100);

    digitalWrite(LED_D0, LOW);
    digitalWrite(LED_D1, LOW);
    digitalWrite(LED_D2, LOW);
    digitalWrite(LED_D3, LOW);
    delay(100);

    }

  //Stop PB check
  if(digitalRead(PIN_A1))
    {
    keep_going = 0;                      //respond to stop PB
```

```
  if(troubleshoot) Serial.println("Stop PB");
  }
delay(2000);                        //2 second delay
if(troubleshoot) Serial.println(" ");
loop_cnt++;
}

//*************************************************************
```

5.6.8 GCS Printed Circuit Board

The layout for the GCS printed circuit board (PCB) and the actual circuit board design are provided in Fig. 5.12. The PCB is mounted to a DIN rail using DIN rail mounting adaptors (Molence C45 35 × 15 mm bracket).

5.6.9 Enclosure

The completed GCS System is mounted within a QILIPSU hinged cover, stainless steel latch, junction box with mounting plate. The layout of the junction box is provided in Fig. 5.13 and the final result is provided in Fig. 5.14.

5.7 Testing

The final project step is to thoroughly test all system features. A test plan is developed to test and document the proper operation of each system feature and the overall system. A test plan for the Greenhouse Control System is provided in Fig. 5.15. Should a test fail, the software is corrected. The test plan is then restarted.

In developing and testing software for a system, it is not always possible or desirable to have close access to the system. For example, in the development of control software for a greenhouse, the greenhouse is not always readily available as shown in Fig. 5.16. Also, different weather conditions are not conveniently available to test the control algorithm under a variety of conditions. In these situations a simulator may be used to substitute for the system. The simulator provides the necessary inputs and signals in place of the system so that software may be developed. I have used the technique when developing control systems for a high end expensive audio amplifier system and also for industrial door controllers.

In this vein, a Greenhouse simulator was developed as shown In Fig. 5.17. For the simulator, potentiometers were used as sensor simulators and LEDs were used for the pump

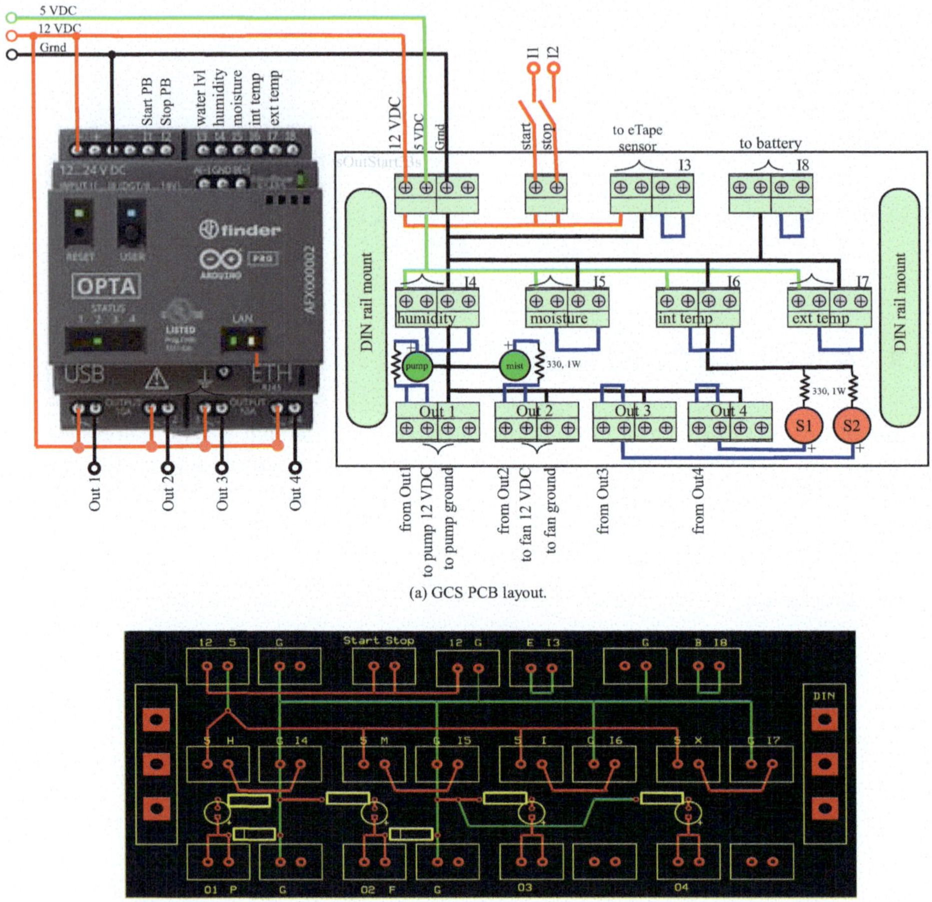

(a) GCS PCB layout.

(b) GCS PCB design.

Fig. 5.12 Greenhouse printed circuit board

and fan. The completed test circuit is shown in Fig. 5.18. The simulator allows exhaustive testing of different weather conditions and control system operation.

5.8 Application: Greenhouse Control System–Ladder Logic

Provided in Fig. 5.19 is the beginning of a Greenhouse Control System in ladder logic. Complete the control system.

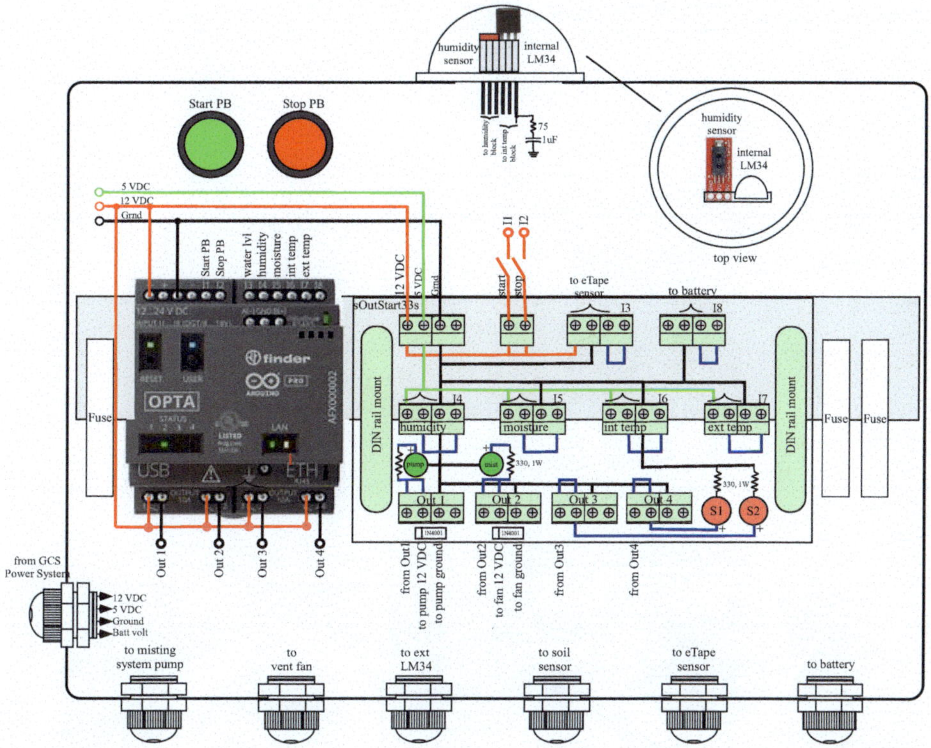

Fig. 5.13 GCS panel layout

5.9 Application: Opta WiFi Bluetooth BLE Greenhouse Monitor

In the following example, we develop a Bluetooth BLE application to gather greenhouse data and make it available for viewing on a client cell phone. As before, we use a greenhouse simulator as a substitute during software development.

In this example we use the Arduino Opta WiFi (AFX00002) as a server for greenhouse parameters and make them available to BLE peripheral clients. A cell phone serves as a client. Through a BLE app (e.g. nRF Connect, LightBlue), the cellphone is used to read greenhouse parameters and to control a simulated vent fan or water pump.

Provided in the sketch below is the Greenhouse Control System. Additions have been made to convert sensor readings to actual values of temperature, humidity, etc. Adapt the example provided in Chap. 2 to monitor greenhouse parameters and control the mist pump and vent fan from a cell phone.

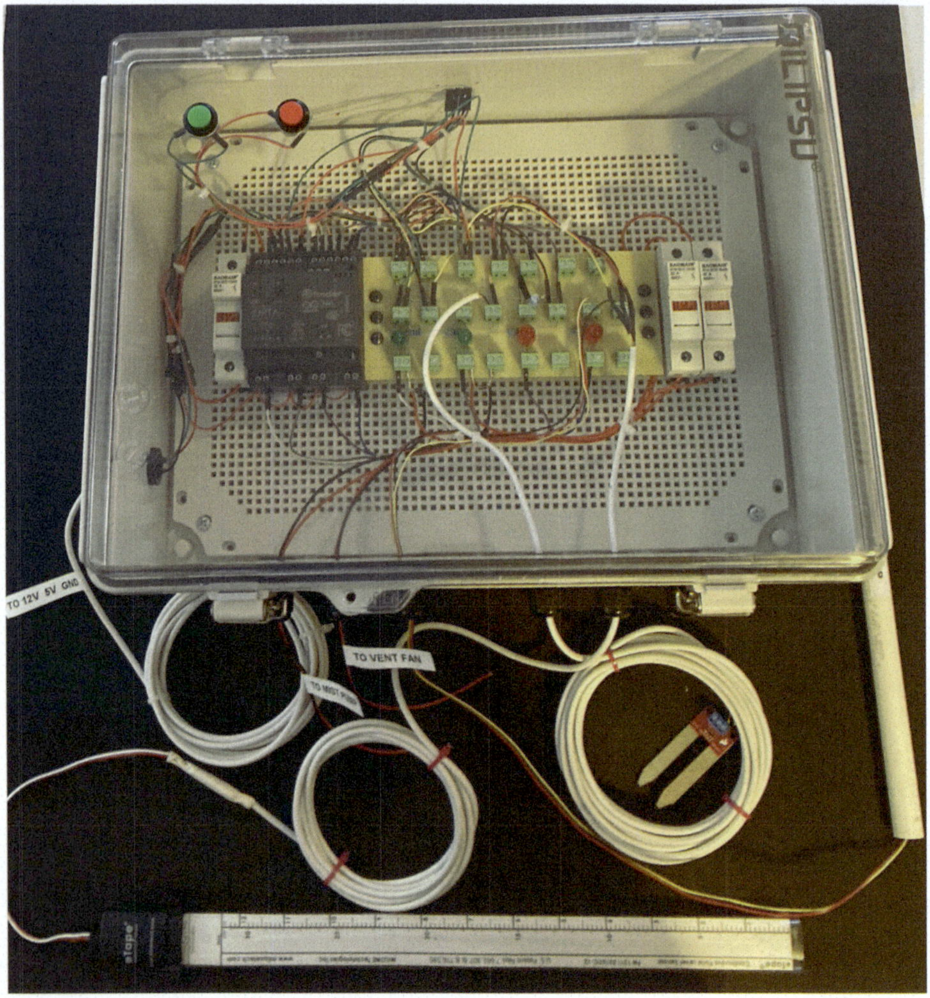

Fig. 5.14 GCS system

```
//********************************************************************
//Opta_GH_control2
//
//Monitors:
//I1: Start - Green PB (digital)
//I2: Stop - Red PB (digital)
//I3: Water level (analog) - Milone e-Tape sensor
//I4: Humidity (analog) - Honeywell HIH sensor
//I5: Soil moisture (analog) - Sparkfun moisture sesnor
//I6: Internal temperature (analog) - LM34
//I7: External temperature (analog) - LM34
//I8: Battery voltage level (analog) - 1/2V from 12V lead acid batt
```

Test #	Description	Expected Status	Results
1	Start program	Program should bypass main loop until Start PB pressed. When Start PB pressed, program should continuously process main loop.	
2	Data collection.	Within main loop, program should **collect data from "sensors"** (potentiometers). Program should respond to changes in **"sensor" values.**	
3	Vent fan	When internal temperature measurement exceeds threshold setting, vent fan and LED will come on. Fan and LED turns off when internal temperature falls below threshold setting.	
4	Mist pump	When soil moisture measurement falls below threshold setting, mist pump and LED will come on. Mist pump and LED turns off when soil moisture is above threshold setting.	
5	Battery status	When battery voltage measurement falls below threshold setting, Status LED 1 will come on. Status LED 1 turns off when battery voltage is above threshold setting.	
6	Stop program	When Stop PB is pressed, program stops **data collection and responding to "sensor"** values changes. Opta status LEDs will sequentially flash.	

Fig. 5.15 GCS test plan

```
//
//Controls:
//Out 1: Water pump for misting system
//Out 2: Vent fan for excessive internal temperature
//
//Status:
//Out 1: Pump LED
//Out 2: Vent fan LED
//Out 3: Status 1 LED
//Out 4: Status 2 LED
//
//Notes:
//- Opta Wi-Fi required for BlueTooth BLE link
//- Thresholds based on 12 bit ADC, Vmax = 10 VDC
//
//Last revised:  02.02.24  S. Barrett
//**************************************************************

unsigned int troubleshoot = 1;      //1: serial monitor status prints
                                    //set status thresholds
unsigned int keep_going = 0;        //loop variable for start/stop
unsigned int water_level_th = 1024; //water level threshold
unsigned int humidity_th = 1024;    //humidity level threshold
```

Fig. 5.16 Greenhouse availability. Mother nature had other plans

```
unsigned int moisture_th = 1024;     //moisture level threshold
unsigned int int_temp_th = 370;      //GH int temp threshold
unsigned int ext_temp_th = 370;      //GH ext temp threshold
unsigned int batt_lvl_th = 2458;     //batt level threshold
                                     //status values
unsigned int water_level;            //water level value
unsigned int humidity;               //humidity level value
unsigned int moisture;               //mositure level value
unsigned int int_temp;               //GH int temp value
unsigned int ext_temp;               //GH ext temp value
unsigned int batt_lvl;               //batt level value
unsigned int loop_cnt =0;            //loop counter

float        actual_water_level;     //water level value
unsigned int actual_humidity;        //humidity level value
unsigned int actual_moisture;        //mositure level value
unsigned int actual_int_temp;        //GH int temp value
unsigned int actual_ext_temp;        //GH ext temp value
unsigned int actual_batt_lvl;        //batt level value

float        int_degrees, ext_degrees;
```

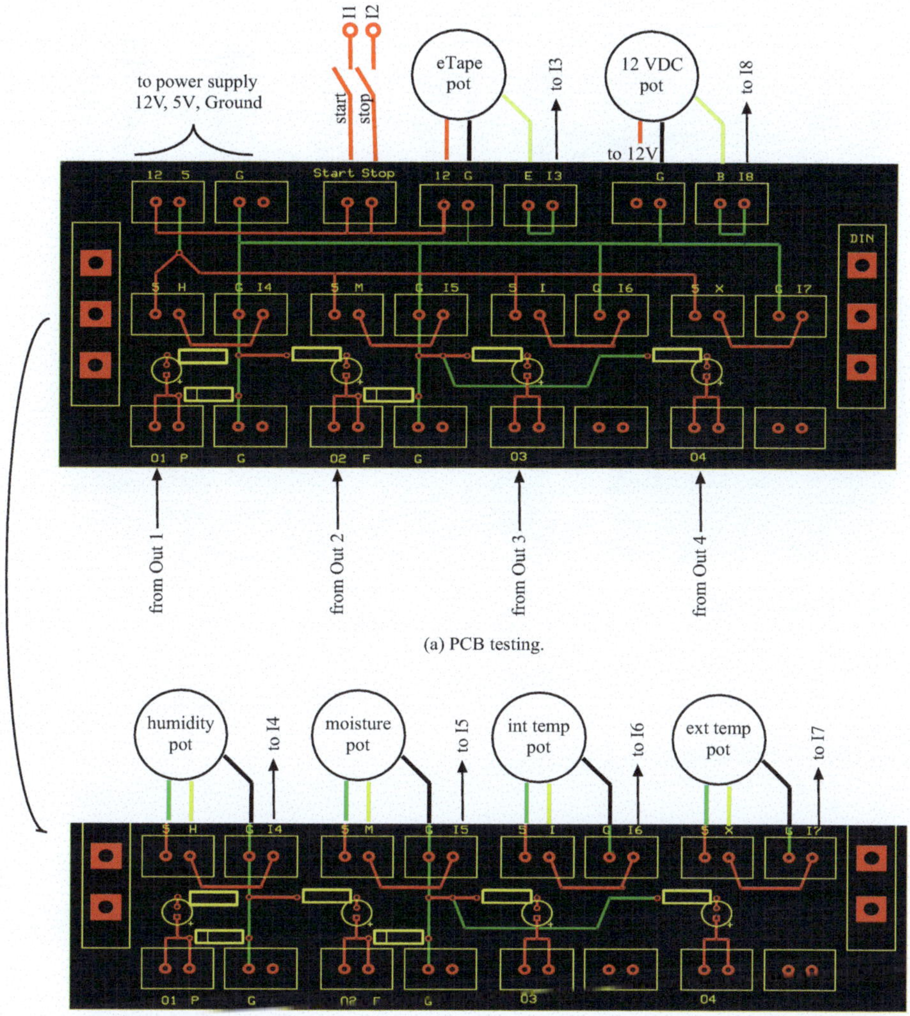

(a) PCB testing.

(b) PCB testing. Connections for second row of sensors.

Fig. 5.17 GCS text fixture layout

```
void setup()
{
if(troubleshoot)Serial.begin(9600); //configure Serial Monitor

analogReadResolution(12);           //ADC set 12 to 16 bits
pinMode(PIN_A0, INPUT);             //Init Start input I1
pinMode(PIN_A1, INPUT);             //Init Stop input I2
```

Fig. 5.18 GCS text fixture

```
pinMode(LED_D0, OUTPUT);                //Initialize status LEDs 1 to 4
pinMode(LED_D1, OUTPUT);
pinMode(LED_D2, OUTPUT);
pinMode(LED_D3, OUTPUT);

pinMode(D0, OUTPUT);                    //Initialize relays outputs
pinMode(D1, OUTPUT);
pinMode(D2, OUTPUT);
pinMode(D3, OUTPUT);
}

void loop()
{
if(troubleshoot) Serial.print("loop begin: ");
if(troubleshoot) Serial.println(loop_cnt);
if((digitalRead(PIN_A0))||(keep_going))//start or keeping going
  {
  keep_going = 1;                       //seal start PB
  if(troubleshoot) Serial.println("keep going");

  //Update status
  water_level = analogRead(A2);         //I3
  humidity    = analogRead(A3);         //I4
  moisture    = analogRead(A4);         //I5
  int_temp    = analogRead(A5);         //I6
  ext_temp    = analogRead(A6);         //I7
  batt_lvl    = analogRead(A7);         //I8
```

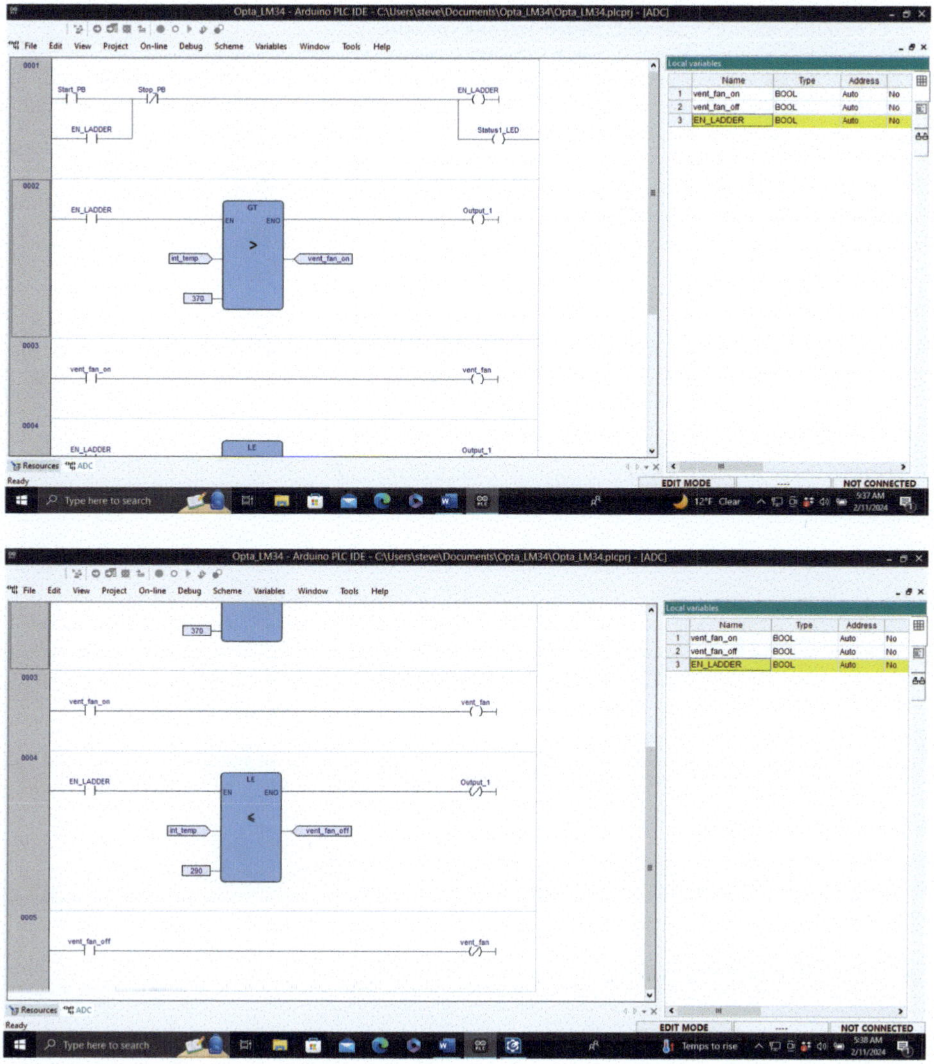

Fig. 5.19 Greenhouse control system in ladder logic

```
if(troubleshoot)
  {
  if(troubleshoot) Serial.print("water level: ");
  if(troubleshoot) Serial.println(water_level);
  if(troubleshoot) Serial.print("humidity: ");
  if(troubleshoot) Serial.println(humidity);
  if(troubleshoot) Serial.print("moisture: ");
  if(troubleshoot) Serial.println(moisture);
  if(troubleshoot) Serial.print("int temp: ");
```

```
  if(troubleshoot) Serial.println(int_temp);
  if(troubleshoot) Serial.print("ext temp: ");
  if(troubleshoot) Serial.println(ext_temp);
  if(troubleshoot) Serial.print("batt lvl: ");
  if(troubleshoot) Serial.println(batt_lvl);
  }

//Update controls and status LEDs

//vent fan
if(int_temp >= int_temp_th)
  {
  digitalWrite(D1, HIGH);        //vent van on
  if(troubleshoot) Serial.println("vent fan on");
  }
else
  {
  digitalWrite(D1, LOW);         //vent van off
  if(troubleshoot) Serial.println("vent fan off");
  }

//soil moisture
if(moisture <= moisture_th)
  {
  digitalWrite(D0, HIGH);        //mist pump on
  if(troubleshoot) Serial.println("mist pump on");
  }
else
  {
  digitalWrite(D0, LOW);         //mist pump off
  if(troubleshoot) Serial.println("mist pump off");
  }

  //battery level
  if(batt_lvl <= batt_lvl_th)
  {
  digitalWrite(D2, HIGH);        //Status LED 1 on
  if(troubleshoot) Serial.println("battery low");
  }
else
  {
  digitalWrite(D2, LOW);         //Status LED 1 off
  if(troubleshoot) Serial.println("battery ok");
  }

//Convert sensor readings to actual values
//I3: water level - determine for actual
                                 //convert to voltage
  actual_water_level = water_level/4096.0 * 10.0;
                                 //5 VDC = 12" fluid
  actual_water_level = actual_water_level/5.0 *12.0;
  if(troubleshoot)
    {
    Serial.print("Water level [in]: ");
```

```
    Serial.println(actual_water_level);
    }

//I4: humidity [%]
//actual_humidity
float humidity_voltage = ((float)(humidity) /4096.0) * 10.0;
                                //convert to RH per data sheet
//float sensor_RH_flt  = ((humidity_voltage/5.0)*161.0) - 25.81;
float sensor_RH_flt  = (((humidity_voltage/5.0) - 0.16) * (1/0.0062));
                                //convert temp reading to C
float int_temp_C = (float)(((float)(actual_int_temp) - 32.0) * (5.0/9.0));
                                //compensate for temp per data sheet
float true_RH = sensor_RH_flt/(1.0546 - (0.00216 * int_temp_C));

if(troubleshoot)
    {
    Serial.print("Humidity: ");
    Serial.println(humidity);
    Serial.print("Humidity voltage: ");
    Serial.println(humidity_voltage);
    Serial.print("sensor RH flt: ");
    Serial.println(sensor_RH_flt);
    Serial.print("Int Temp F: ");
    Serial.println(actual_int_temp);
    Serial.print("Int Temp C: ");
    Serial.println(int_temp_C);
    Serial.print("Relative Humidity: ");
    Serial.println(true_RH);
    }

//I5: moisture - determine for actual configuration
actual_moisture = moisture;
if(troubleshoot)
  {
  Serial.print("soil moisture: ");
  Serial.println(actual_moisture);
  }

//I6: interior temp degrees F
float voltageI6 = (float)(int_temp) * (10.0/4095.0);
int_degrees = voltageI6 * 100.0;
actual_int_temp = (unsigned int)(int_degrees);
if(troubleshoot)
  {
  Serial.print("I6 value: ");         //Print out value from I6
  Serial.print(voltageI6, 2);         //voltage two digits
  Serial.println(" Volts");
  Serial.print("Int Temp:");
  Serial.print(" ");
  Serial.print(int_degrees);
  Serial.println(" degrees (F)");
  Serial.print("Int Temp:");
  Serial.print(" ");
  Serial.print(actual_int_temp);
```

```
    Serial.println(" degrees (F)");
    }

  //I7: exterior temp degrees F

  //I8: battery level % of 12 VDC
  actual_batt_lvl = (batt_lvl * 100)/batt_lvl_th;     //I8
  if(troubleshoot)
    {
    Serial.print("Battery [%]: ");
    Serial.println(actual_batt_lvl);
    }

  //Update BLE status report

  }
else
  {
  if(troubleshoot) Serial.println("keep going - no");
  keep_going = 0;                    //stop
  //flash Opta panel LEDs
  digitalWrite(LED_D0, HIGH);
  delay(100);
  digitalWrite(LED_D0, LOW);
  delay(100);

  digitalWrite(LED_D1, HIGH);
  delay(100);
  digitalWrite(LED_D1, LOW);
  delay(100);

  digitalWrite(LED_D2, HIGH);
  delay(100);
  digitalWrite(LED_D2, LOW);
  delay(100);

  digitalWrite(LED_D3, HIGH);
  delay(100);
  digitalWrite(LED_D3, LOW);
  delay(100);

  digitalWrite(LED_D0, LOW);
  digitalWrite(LED_D1, LOW);
  digitalWrite(LED_D2, LOW);
  digitalWrite(LED_D3, LOW);
  delay(100);

  }

//Stop PB check
if(digitalRead(PIN_A1))
  {
  keep_going = 0;                    //respond to stop PB
  if(troubleshoot) Serial.println("Stop PB");
```

```
  }
delay(2000);                    //2 second delay
if(troubleshoot) Serial.println(" ");
loop_cnt++;
}

//*************************************************************
```

5.10 Summary

The goal of this chapter was to demonstrate in action the concepts discussed in this book. Simply put, our goal is to provide the theory, design, and construction of a passively heated greenhouse. We equip the greenhouse with instrumentation to monitor and control key parameters. Using IoT concepts key parameters will be monitored and controlled via a remote computer.

5.11 Problems

1. Provide a design to allow the LM34 to sense negative temperatures.
2. Provide a detailed structure chart for the GCS system code.
3. Provide the overall GCS system code.
4. Develop the overall system software for the GCS System.
5. Develop a test plan to insure requirements have been met for the GCS system.
6. What are the critical variables that should be regularly monitored?
7. If the greenhouse is within 10 m of your home, design a system to report critical variables to your home PC. Explain in detail your choice of technology.
8. If the greenhouse is within 10 m of your home, design a system to report critical variables to your cell phone. Explain in detail your choice of technology.
9. If the greenhouse is within 50 m of your home, design a system to report critical variables to your home PC. Explain in detail your choice of technology.
10. Design a system to log and plot the internal and external greenhouse temperature every hour.

References

1. Arduino homepage, www.arduino.cc
2. Baird C. (2011) *The Complete Guide to Building Your Own Greenhouse*, Atlantic Publishing.
3. Doxon, L. *How to Calculate Greenhouse Heating*, https://homeguides.sfgate.com.
4. Encinias, V. *Efficient Greenhouse Design*, https://gpnmag.com.

5. *Five Low–Tech Winter Greenhouse Heating Techniques*, https://www.rimolgreenhouses.com.
6. Hanes D., G. Salgueiro, P. Grossetete, R. Barton, J. Henry (2017) *IoT Fundamentals–Networking Technologies, Protocols, and Use Cases for the Internet of Things*, Cisco Press.
7. Honeywell, *HIH–4030/31 Series Humidity Sensors*, Honeywell Sensing and Control, www.honeywell.com/sensing.
8. Kelley, A. and I. Pohl (1998), *A Book on C – Programming in C*, 4 th edition, Addison Wesley.
9. Lindsey, C. and M. Plinke (2016) *The Year–Round Solar Greenhouse*, New Society Publishers.
10. Marshall, R. (2006) *How to Build Your Own Greenhouse*, Storey Publishing.
11. Milone Technologies, *0–5 VDC Linear Resistance to Voltage Module PN–05V00199 Rev 2*, www.milonetech.com.
12. Milone Technologies, *eTape Continuous Fluid Level Sensor PN–12110215TC–X*, www.milonetech.com.
13. National Semiconductor, *LM34 Precision Fahrenheit Temperature Sensors*, DS006685, National Semiconductor Corporation, www.national.com, 2000.
14. Oehler, M. (2007) *The Earth–Sheltered Solar Greenhouse Book*, Mole Publishing Company.
15. *Passive Solar Greenhouse*, Bradford Research Center, University of Missouri, 2017, http://bradford.cafnr.org/passive---solar---greenhouse.
16. Schiller, L. *How to Design a Year–Round Solar Greenhouse*, https://www.motherearthnews.com.
17. Schiller, L. *Three Methods for Heating Greenhouses for Free*, https://www.motherearthnews.com.
18. Schmidt, P. (2011) *The Complete Guide to Greenhouses & Garden Projects*, Creative Publishing International.
19. Thoma, M. *Seven Useful Features You Need in a Passive Solar Greenhouse*, tranquilurbanhomestead.com.
20. Watson, G. *Rain Barrels–A Homeowner's Guide,* SW Florida Water Management District, www.WaterMatters.org.

6 Opta Expansions

Objectives: After reading this chapter, the reader should be able to do the following:

- Describe the function of Opta expansions;
- Summarize the features of the Opta expansions;
- Configure an Arduino Opta with appropriate expansions;
- Develop sketches for the Opta expansions using the Arduino IDE; and
- Develop applications to showcase the unique features of the Opta expansions.

6.1 Overview

In my humble opinion, the Arduino company has been highly successful in their ongoing efforts to make complex technology available to all. New developments are being released almost continuously over an extended period of time. As evidence, the recent release of Opta expansions has extended the features of an already powerful line of programmable logic controllers. In this chapter we highlight the unique features provided by the expansions.[1]

6.2 Opta Expansions

Arduino has recently released a series of hardware expansion modules to extend and enhance the Opta PLC capabilities. Provided in Fig. 6.1 is a summary of Opta expansion features. There are currently three expansions available:

[1] I am thankful to Chuck Glaser, Editor, and Prasanna Kumar Narayanasamy, Production Editor, who paused the production schedule to allow inclusion of this chapter.

S. F. Barrett, *Arduino VII*, Synthesis Lectures on Digital Circuits & Systems,
https://doi.org/10.1007/978-3-031-68609-2_6

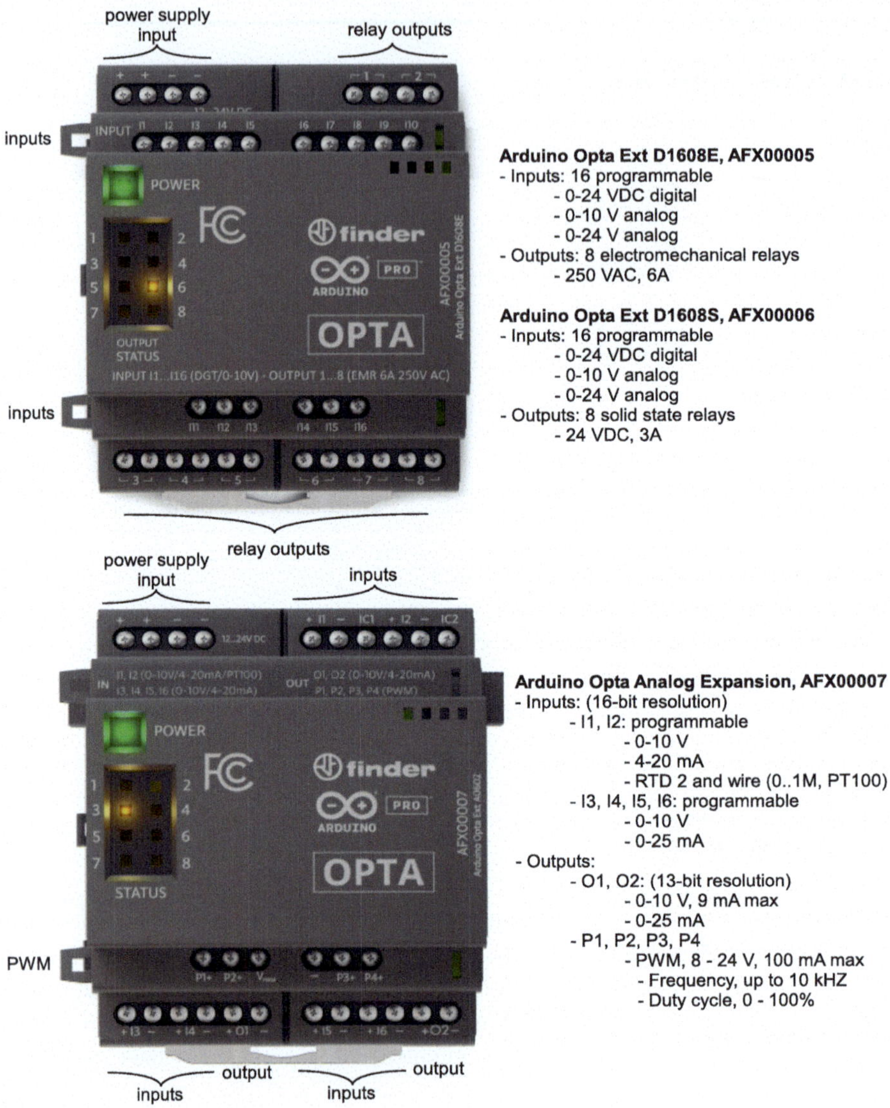

Fig. 6.1 Opta expansion features. Images used courtesy of the Arduino team (CC BY–NC–SA) (www.arduino.cc)

- Arduino Opta Digital Expansion D1608E, AFX00005
- Arduino Opta Digital Expansion D1608S, AFX00006
- Arduino Opta Analog Expansion A0602, AFX00007.

6.2.1 Digital Expansions D1608E and D1608S

The D1608E and D1608S have 16 programmable inputs and eight relay outputs. The D1608E expansion is equipped with electromechanical relays rated at 250 VAC and 6 amps. The relay response time when transitioning from logic 0 to 1 is 5 ms and 3 ms when transitioning from logic 1 to 0.

The D1608S expansion is equipped with solid state relays (SSR) rated at 24 VDC and 3 amps. The relay response time when transitioning from logic 0 to 1 is 0.02 ms and 0.2 ms when transitioning from logic 1 to 0.

6.2.2 Analog Expansion A0602

The A0602 analog expansion is equipped with the following features:

- Inputs I1 and I2 which are programmable for:
  - 0–10 V
  - 4–20 mA
  - Resistance temperature detectors (RTD), both 2–wire and 3–wire
- Inputs I3 to I6 which are programmable for:
  - 0–10 V
  - 4–20 mA
- Outputs O1 and O2 (and I1–I6)[2]:
  - 0–10 V: The maximum output current is 9 mA.
  - 4–20 mA
- Outputs P1–P4 that are configured as pulse width modulations outputs. The PWM voltage (V_{PWM}) must be provided to the A0602 expansion. The PWM frequency and duty cycle (0–100%) is fully programmable. The output current limit for each PWM channel is 100 mA.

[2] Analog inputs can also be programmed as output channels.

6.3 Getting Started

In this section we describe required hardware and software configurations.

6.3.1 Hardware Configuration

The expansion modules are snapped together as shown in Fig. 6.2. The small "AUX" access panel located on the right side of the Opta PLC and the expansions must be carefully removed. The units are then connected via the "AUX connector." The Opta PLC and the expansions conveniently mount to a DIN rail. Up to five expansions may be connected to an Opta PLC. Once connected, power must be provided to each expansion as shown in Fig. 6.2. The supply voltage may be 12–24 VDC.

Fig. 6.2 Opta expansion connections. Images used courtesy of the Arduino team (CC BY–NC–SA) (www.arduino.cc)

6.3.1.1 Opta Expansion Communication

The Opta PLC communicates with the expansions via the Inter–Integrated Circuit (I2C) subsystem aboard the Opta host controller. The I2C connections between the Opta PLC and the expansions are made via the "AUX" connector. The I2C subsystem allows the system designer to connect a number of I2C configured devices together into a system using a two–wire interconnecting scheme. Each device has its own unique address and may both transmit and receive over the two–wire bus at frequencies up to 400 kHz. This allows the device to freely exchange information with other devices in a small area network. The I2C is alternately known as the Two Wire Interface (TWI) protocol (Philips).

6.3.2 Software Configuration

There are a number of examples available in the Arduino IDE "Arduino_Opta_Blueprint" library. The library is available for download via the Library Manager. Once the library is downloaded, upload and execute the sketch "UpdateExpansionFW" and follow the instructions provided via the Serial Monitor. The sketch detects each connected expansion and individually updates their resident firmware. You will be prompted to update each expansion.

6.4 Arduino Opta Blueprint Library

In this section we work through a number of examples provided in the "Arduino_Opta_Blueprint" library. Along the way we provide additional background material to enhance your understanding of Opta expansion operation.

6.4.1 GetExpansion

The "getExpansion" sketch polls each expansion connected to the Opta in turn and reports on the type and I2C address of each expansion. In the example the sketch is applied to the Opta expansion configuration provided in Fig. 6.2.

```
//*************************************************************************************
//FILENAME: getExpansions.ino
//AUTHOR:   Daniele Aimo
//DATE:     20231020
//REVISION: 0.0.1
//DESCRIPTION: Retrieve list of expansions chained to Opta Controller and their type
//PRODUCT:  Arduino Opta
//LICENSE:  Copyright (c) 2024 Arduino SA
//          This Source Code Form is subject to the terms fo the Mozilla Public
//          License (MPL), v 2.0. You can obtain a copy of the MPL
//          at \url{http://mozilla.org/MPL/2.0/.}
//Modified: S. Barrett, Nov 2024
//*************************************************************************************

#include "OptaBlue.h"
```

```
void setup()
{
Serial.begin(115200);
delay(2000);
OptaController.begin();
while (!Serial)                              //wait for serial monitor
  {
  ;
  }
}

void loop()
{
OptaController.update();                     //determines Opta expansion configuration
Serial.print("Number of expansions: ");      //get the number of connected expansions
Serial.println(OptaController.getExpansionNum());

//for each expansion get and print the type and the I2C address of each expansion
for(int i = 0; i < OptaController.getExpansionNum(); i++)
  {
  Serial.print("Expansion n. ");
  Serial.print(i);
  Serial.print(" type ");
  printExpansionType(OptaController.getExpansionType(i));
  Serial.print(" I2C address ");
  Serial.println(OptaController.getExpansionI2Caddress(i));
  }
Serial.println(" ");
delay(2000);
}

//*************************************************************************************

void printExpansionType(ExpansionType_t t)
{
if(t == EXPANSION_NOT_VALID)
  {
  Serial.print("Unknown! (Might be an Unregistered custom expansion?)");
  }
else if(t == EXPANSION_OPTA_DIGITAL_MEC)
  {
  Serial.print("Opta --- DIGITAL [Mechanical]  ---");
  }
else if(t == EXPANSION_OPTA_DIGITAL_STS)
  {
  Serial.print("Opta --- DIGITAL [Solid State] ---");
  }
else if(t == EXPANSION_DIGITAL_INVALID)
  {
  Serial.print("Opta --- DIGITAL [!!Invalid!!] ---");
  }
else if(t == EXPANSION_OPTA_ANALOG)
  {
  Serial.print("Opta ~~~ ANALOG ~~~ ");
  }
else
  {
  Serial.print("Unknown!");
  }
}

//*************************************************************************************
```

Figure 6.3 provides the information retrieved from polling the Opta expansion configuration.

6.4.2 SetDigital

The "setDigital" sketch demonstrates how to set digital outputs on a specific expansion. The sketch determines the number and type of expansions connected to the Opta PLC. The

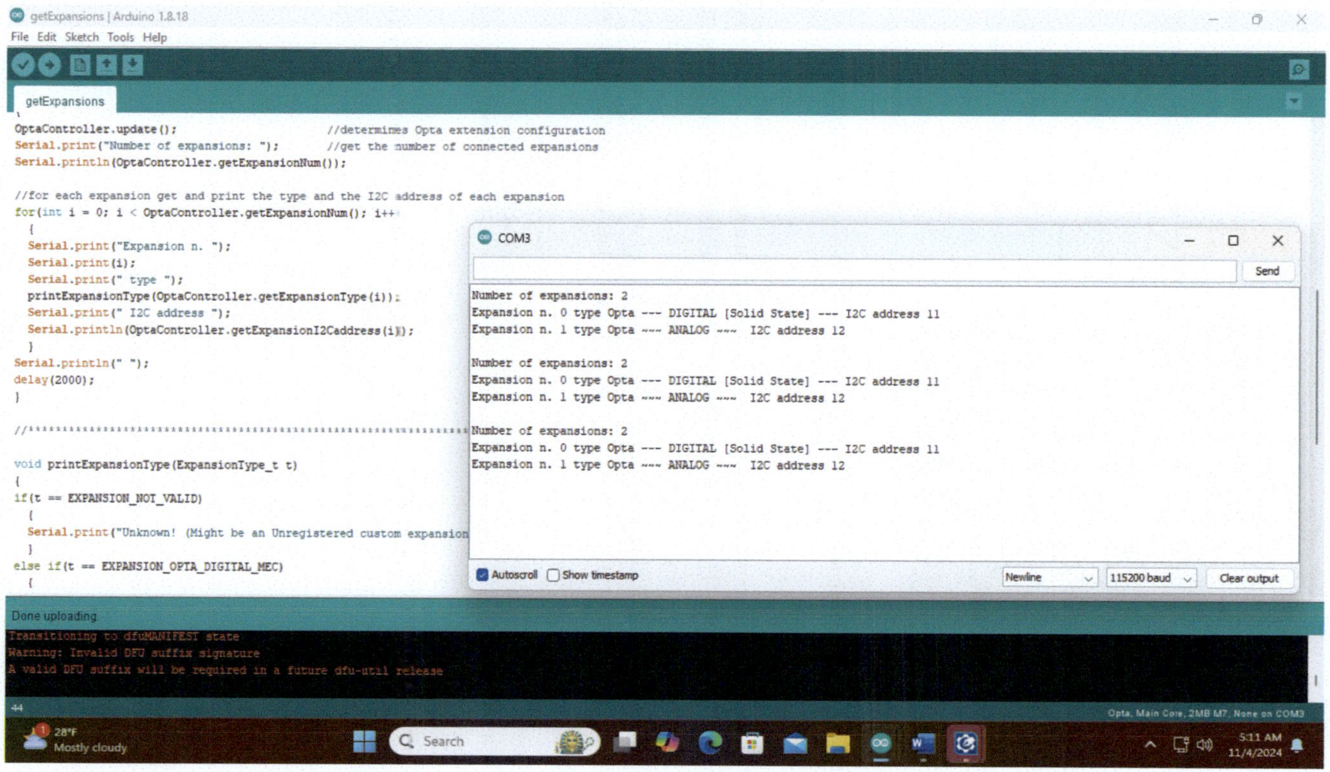

Fig. 6.3 Polling the Opta expansion connections. Images used courtesy of the Arduino team (CC BY–NC–SA) (www.arduino.cc)

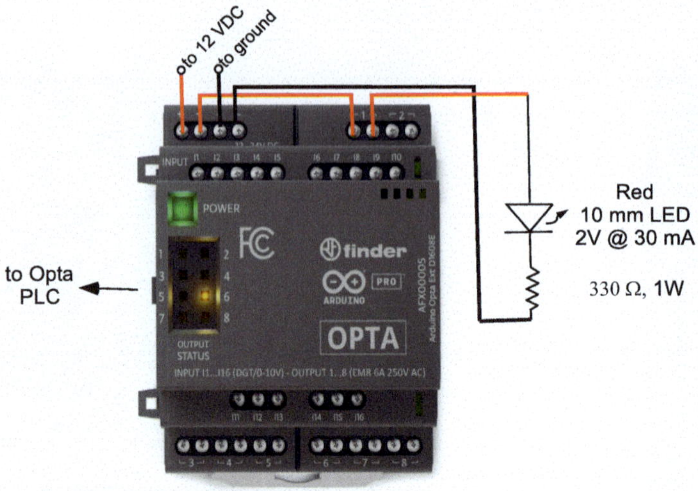

Fig. 6.4 Controlling a load with setDigital sketch (CC BY–NC–SA) (www.arduino.cc)

sketch then blinks LEDs associated with the digital outputs in a specific sequence associated with the specific expansion. The expansion closes a “switch” between the output terminals associated with the LEDs (e.g. O1, O2, etc.) which allows the LED to illuminate. The “switch” is either a mechanical relay (Expansion D1608E) or a solid state relay (Expansion D1608S). The configuration for connecting an external load (e.g. an LED) is shown in Fig. 6.4.

```
//*****************************************************************************
//FILENAME: setDigital.ino
//AUTHOR:   Daniele Aimo
//EMAIL:    d.aimo@arduino.cc
//DATE:     20231116
//DESCRIPTION: Sketch shows how to set the digital values of Digital
//             Expansion digital output pins
//LICENSE:  Copyright (c) 2024 Arduino SA
//          This Source Code Form is subject to the terms fo the Mozilla
//          Public License (MPL), v 2.0. You can obtain a copy of the MPL
//          at \url{http://mozilla.org/MPL/2.0/.}
//Modified: S. Barrett, Nov 2024
//*****************************************************************************

#include "OptaBlue.h"

using namespace Opta;

void setup()
{
Serial.begin(115200);
delay(2000);
OptaController.begin();
while (!Serial)                          //wait for serial monitor
```

```
  {
  ;
  }
}

void loop()
{
OptaController.update();

static long int start_m = millis();
static bool st = true;

if(millis() - start_m > 1000)                //loops every 1s (1000 ms)
{
start_m = millis();

for(int i = 0; i < OPTA_CONTROLLER_MAX_EXPANSION_NUM; i++)
  {
  DigitalMechExpansion mechExp = OptaController.getExpansion(i);
  DigitalStSolidExpansion stsolidExp = OptaController.getExpansion(i);
  if(mechExp)                //Expansion with mechanical relays
    {                        //get and print information about expansion
    printExpansionInfo(mechExp.getIndex(), mechExp.getType(),
    mechExp.getI2CAddress());
                            //implements two states
                            //state 1: pin 0 2 4 6 are turned off
                            //and pin 1 3 5 7 are turned on
    if(st)
      {
      mechExp.digitalWrite(0, LOW);      //turn off pin 0
      mechExp.digitalWrite(1,HIGH);      //turn on pin 1
      mechExp.digitalWrite(2, LOW);      //turn off pin 2
      mechExp.digitalWrite(3,HIGH);      //turn on pin 3
      mechExp.digitalWrite(4, LOW);      //turn off pin 4
      mechExp.digitalWrite(5,HIGH);      //turn on pin 5
      mechExp.digitalWrite(6, LOW);      //turn off pin 6
      mechExp.digitalWrite(7,HIGH);      //turn on pin 7

      //once all pin are set, send new status to expansion
      mechExp.updateDigitalOutputs();
      }
    else
      {
                           //state 2: pin 0 2 4 6 are turned on
                           //and pin 1 3 5 7 are turned off
      mechExp.digitalWrite(0,HIGH);      //turn off pin 0
      mechExp.digitalWrite(1, LOW);      //turn on pin 1
      mechExp.digitalWrite(2,HIGH);      //turn off pin 2
      mechExp.digitalWrite(3, LOW);      //turn on pin 3
      mechExp.digitalWrite(4,HIGH);      //turn off pin 4
      mechExp.digitalWrite(5, LOW);      //turn on pin 5
      mechExp.digitalWrite(6,HIGH);      //turn off pin 6
      mechExp.digitalWrite(7, LOW);      //turn on pin 7

      //once all pin are set, send new status to expansion
      mechExp.updateDigitalOutputs();
      }//end else
  }//end (if mechExp)

if(stsolidExp)              //Expansion with solid state relays
  {
  printExpansionInfo(stsolidExp.getIndex(), stsolidExp.getType(),
```

```
  stsolidExp.getI2CAddress());
                            //if present state solid expansion
                            //will use a different pattern
  if(st)
    {
    stsolidExp.digitalWrite(0,HIGH);
    stsolidExp.digitalWrite(1, LOW);
    stsolidExp.digitalWrite(2, LOW);
    stsolidExp.digitalWrite(3,HIGH);
    stsolidExp.digitalWrite(4,HIGH);
    stsolidExp.digitalWrite(5, LOW);
    stsolidExp.digitalWrite(6, LOW);
    stsolidExp.digitalWrite(7,HIGH);

    //once all pin are set, send the new status to the expansion
    stsolidExp.updateDigitalOutputs();
    }
    else
    {
                             //in the second state
                             //pin 0 2 4 6 are turned on
                             //and pin 1 3 5 7 are turned off
    stsolidExp.digitalWrite(0,LOW);
    stsolidExp.digitalWrite(1,HIGH);
    stsolidExp.digitalWrite(2,HIGH);
    stsolidExp.digitalWrite(3,LOW);
    stsolidExp.digitalWrite(4,LOW);
    stsolidExp.digitalWrite(5,HIGH);
    stsolidExp.digitalWrite(6,HIGH);
    stsolidExp.digitalWrite(7,LOW);

    //once all pin are set send the new status to the expansion
    stsolidExp.updateDigitalOutputs();
    }
  }
}
if(st)
  {
  st = false;
  }
else
  {
  st = true;
  }
 }
}

//********************************************************************************

void printExpansionType(ExpansionType_t t)
{
if(Serial)
  {
  if(t == EXPANSION_NOT_VALID)
    {
    Serial.print("Unknown!");
    }
  else if(t == EXPANSION_OPTA_DIGITAL_MEC)
    {
    Serial.print("DIGITAL [Mechanical]");
    }
  else if(t == EXPANSION_OPTA_DIGITAL_STS)
```

```
    {
    Serial.print("DIGITAL [Solid State]");
    }
  else if(t == EXPANSION_DIGITAL_INVALID)
    {
      Serial.print("DIGITAL [!!Invalid!!]");
    }
  else if(t == EXPANSION_OPTA_ANALOG)
    {
    Serial.print("ANALOG");
    }
  else
    {
    Serial.print("Unknown!");
    }
  }
}

//*******************************************************************************

void printExpansionInfo(uint8_t index, ExpansionType_t type, uint8_t i2c_address)
{
if(Serial)
  {
  Serial.print("Expansion[" + String(index) + "]:");
  Serial.print(" type ");
  printExpansionType(type);
  Serial.print(", I2C address: ");
  Serial.println(i2c_address);
  }
}

//*******************************************************************************
```

6.4.3 GetDigital

The "getDigital" sketch demonstrates how to read digital inputs on a specific expansion. The sketch determines the number and type of expansions connected to the Opta PLC. The sketch then reads the inputs from each expansion and reports whether they are logic high (H) or logic low (L). In this example a momentary contact pushbutton switch is connected to I6 as shown in Fig. 6.5.

```
//*******************************************************************************
//FILE NAME:    getDigital.ino
//AUTHOR:       Daniele Aimo
//EMAIL:        d.aimo@arduino.cc
//DATE:         20231116
//DESCRIPTION: This sketch shows how to get the digital values of Digital
//              Expansion digital input pins
//LICENSE:      Copyright (c) 2024 Arduino SA
//              This Source Code Form is subject to the terms fo the Mozilla
//              Public License (MPL), v 2.0. You can obtain a copy of the MPL
//              at \url{http://mozilla.org/MPL/2.0/.}
//Modified:     S. Barrett, Nov 2024                                        */
//*******************************************************************************
```

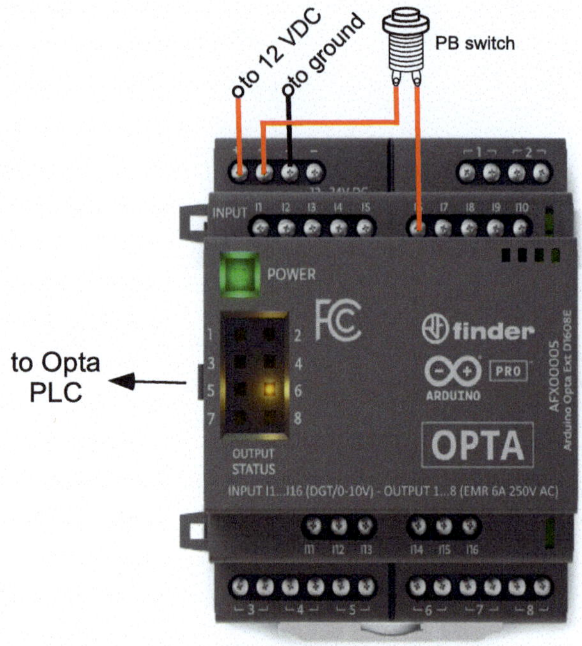

Fig. 6.5 Test circuit for getDigital sketch (CC BY–NC–SA) (www.arduino.cc)

```
#include "OptaBlue.h"

using namespace Opta;

void setup()
{
Serial.begin(115200);
delay(2000);
OptaController.begin();
while (!Serial)                                //wait for serial monitor
  {
  ;
  }
}

void loop()
{
OptaController.update();
Serial.println();
for(int i = 0; i < OPTA_CONTROLLER_MAX_EXPANSION_NUM; i++)
{
DigitalMechExpansion mechExp = OptaController.getExpansion(i);
DigitalStSolidExpansion stsolidExp = OptaController.getExpansion(i);

if(mechExp)
  {
  printExpansionInfo(mechExp.getIndex(), mechExp.getType(),
  mechExp.getI2CAddress());
  mechExp.updateDigitalInputs();

  for(int k = 0; k < OPTA_DIGITAL_IN_NUM; k++)
```

```
    {
    PinStatus v = mechExp.digitalRead(k);     //returns pin status of pin k
    if(v == HIGH)
      {
      Serial.print("H");
      }
    else
      {
      Serial.print("L");
      }
    Serial.print(' ');
    }
 Serial.println();
 }

if(stsolidExp)
  {
  printExpansionInfo(stsolidExp.getIndex(), stsolidExp.getType(),
  stsolidExp.getI2CAddress());
  stsolidExp.updateDigitalInputs();

  for(int k = 0; k < OPTA_DIGITAL_IN_NUM; k++)  //returns pin status of pin k
    {
    PinStatus v = stsolidExp.digitalRead(k);
    if(v == HIGH)
      {
      Serial.print("H");
      }
    else
      {
      Serial.print("L");
      }
    Serial.print(' ');
    }
  Serial.println();
  }
}
delay(1000);
}

//**********************************************************************************

void printExpansionType(ExpansionType_t t)
{
if(t == EXPANSION_NOT_VALID)
  {
  Serial.print("Unknown!");
  }
else if(t == EXPANSION_OPTA_DIGITAL_MEC)
  {
  Serial.print("DIGITAL [Mechanical]");
  }
else if(t == EXPANSION_OPTA_DIGITAL_STS)
  {
  Serial.print("DIGITAL [Solid State]");
  }
else if(t == EXPANSION_DIGITAL_INVALID)
  {
  Serial.print("DIGITAL [!!Invalid!!]");
  }
else if(t == EXPANSION_OPTA_ANALOG)
  {
```

```
  Serial.print("ANALOG");
  }
else
  {
  Serial.print("Unknown!");
  }
}

//******************************************************************************

void printExpansionInfo(uint8_t index, ExpansionType_t type, uint8_t i2c_address)
{
Serial.print("Expansion[" + String(index) + "]:");
Serial.print(" type ");
printExpansionType(type);
Serial.print(", I2C address: ");
Serial.println(i2c_address);
}

//******************************************************************************

void printUint16(uint16_t v)
{
if(v < 10)
  {
  Serial.print("    ");
  }
else if(v < 100)
  {
  Serial.print("   ");
  }
else if(v < 1000)
  {
  Serial.print("  ");
  }
else if(v < 10000)
  {
  Serial.print(" ");
  }
Serial.print(v);
}

//******************************************************************************
```

The switch connected to I6 is pressed momentarily and released. The resulting logic high (H) signal is reported via the Serial Monitor. The results of testing the sketch is provided in Fig. 6.6.

6.4.4 ADC

The sketch “ADC” is used to test the analog input of the Arduino Opta Analog Expansion A0602. This expansion module is equipped with six analog programmable inputs with 16–bit ADC resolution.

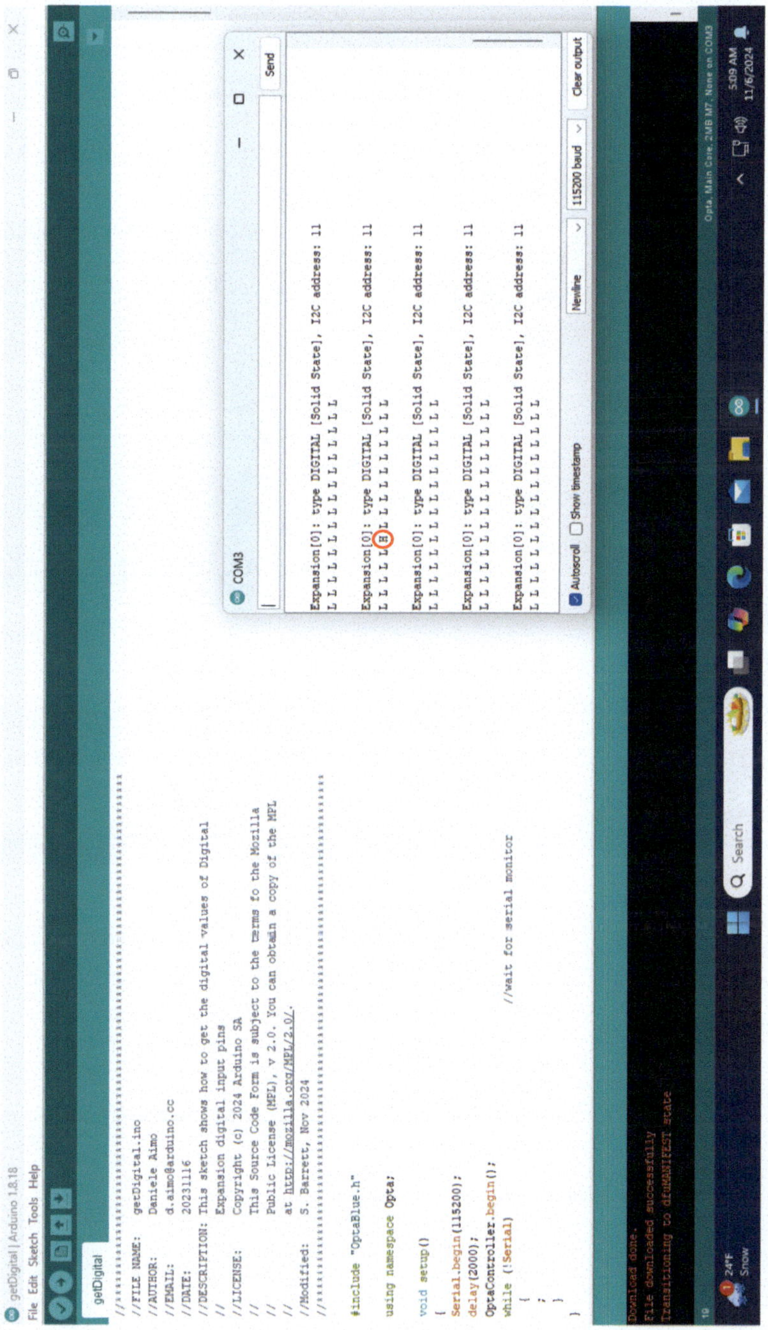

Fig. 6.6 Results of testing getDigital sketch (CC BY–NC–SA) (www.arduino.cc)

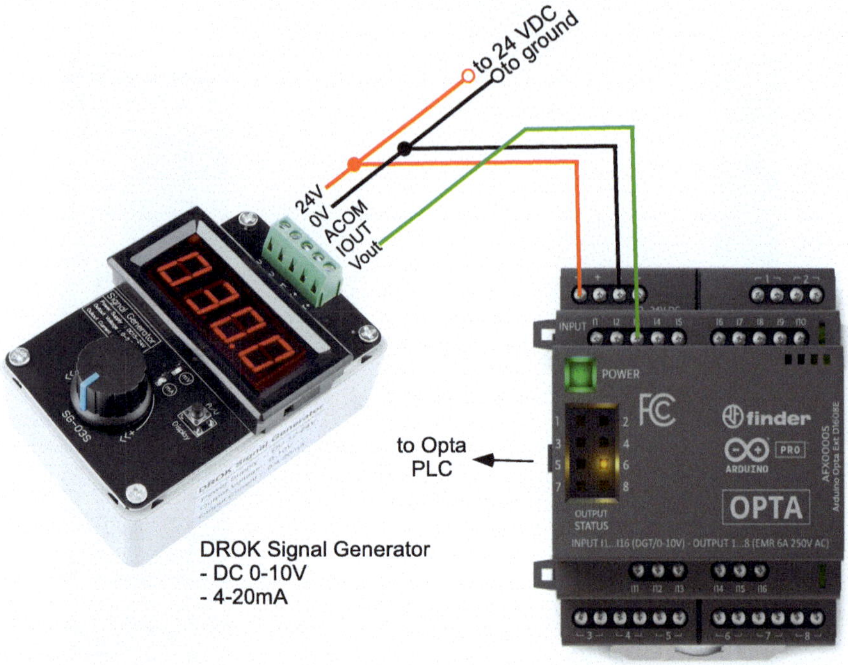

Fig. 6.7 DROK SG–03 signal generator (CC BY–NC–SA) (www.arduino.cc)

We use a DROK SG–03 Signal Generator to provide a 0–10 VDC or a 4–20 mA signal to the analog expansion inputs as shown in Fig. 6.7.

```
//*****************************************************************************
//FILE NAME:    ADC.ino
//AUTHOR:       Daniele Aimo
//DATE:         20231211
//DESCRIPTION: This sketch shows how to use Opta Analog ADC channel
//              This will work only on Opta Analog Expansions
//LICENSE:      Copyright (c) 2024 Arduino SA
//              This Source Code Form is subject to the terms of the Mozilla
//              Public License (MPL), v 2.0. You can obtain a copy of the MPL
//              at \url{http://mozilla.org/MPL/2.0/.}
//Modified:     S. Barrett, Nov 2024                                          */
//*****************************************************************************

#include "OptaBlue.h"
#define PERIODIC_UPDATE_TIME 500
#define DELAY_AFTER_SETUP 5000

using namespace Opta;

int8_t oa_index = -1;

void setup()
{
Serial.begin(115200);
```

```
delay(2000);
OptaController.begin();
while (!Serial)                                    //wait for serial monitor
  {
  ;
  }

for(int i = 0; i < OptaController.getExpansionNum(); i++)
  {
  for(int k = 0; k < OA_AN_CHANNELS_NUM;k++)
    {
    //all input channels initialized as VOLTAGE ADC
    AnalogExpansion::beginChannelAsAdc(OptaController,
          i,                //the device
          k,                //specific output channel
          OA_VOLTAGE_ADC,   //ADC type
          true,             //enable pull down
          false,            //disable rejection
          false,            //disable diagnostic
          0);               //disable averaging
    }
  }
}


void loop()
{
OptaController.update();
optaAnalogTask();
}

//*******************************************************************************

void printExpansionType(ExpansionType_t t)
{
if(t == EXPANSION_NOT_VALID)
  {
  Serial.print("Unknown!");
  }
else if(t == EXPANSION_OPTA_DIGITAL_MEC)
  {
  Serial.print("Opta --- DIGITAL [Mechanical]  ---");
  }
else if(t == EXPANSION_OPTA_DIGITAL_STS)
  {
  Serial.print("Opta --- DIGITAL [Solid State] ---");
  }
else if(t == EXPANSION_DIGITAL_INVALID)
  {
  Serial.print("Opta --- DIGITAL [!!Invalid!!] ---");
  }
else if(t == EXPANSION_OPTA_ANALOG)
  {
  Serial.print("~~~ Opta  ANALOG ~~~");
  }
else
  {
  Serial.print("Unknown!");
  }
}

//*******************************************************************************
```

```
void printExpansionInfo()
{
static long int start = millis();

if(millis() - start > 5000)
  {
  start = millis();
  Serial.print("Number of expansions: ");
  Serial.println(OptaController.getExpansionNum());

  for(int i = 0; i < OptaController.getExpansionNum(); i++)
    {
    Serial.print("Expansion n. ");
    Serial.print(i);
    Serial.print(" type ");
    printExpansionType(OptaController.getExpansionType(i));
    Serial.print(" I2C address ");
    Serial.println(OptaController.getExpansionI2Caddress(i));
    }
  }
}

//*****************************************************************************

void optaAnalogTask()
{
static long int start = millis();
if(millis() - start > PERIODIC_UPDATE_TIME)
  {
  start = millis();
  for(int i = 0; i < OptaController.getExpansionNum(); i++)
    {
    AnalogExpansion exp = OptaController.getExpansion(i);
    if(exp)
      {
        Serial.println("\nAnalog Expansion n. " +  String(exp.getIndex()));

      for(int j = 0; j < OA_AN_CHANNELS_NUM; j++)
        {
        Serial.print(" - ch " + String(j));
          int value =   exp.analogRead((uint8_t)j);
          Serial.println(" -> ADC " + String(value));

        //convert to voltage - 16 bit, 10V max
        float ADC_voltage =(((float)(value))/65535.0) * 10.0;
        Serial.print("CH voltage:  ");
        Serial.println(ADC_voltage);
        Serial.println();
          }
      Serial.println();
      }
    }
  }
}

//*****************************************************************************
```

The results of testing the sketch is provided in Fig. 6.8. The DROK signal generator was connected to channel 2. The ADC reading was converted to an analog result as shown in the sketch above.

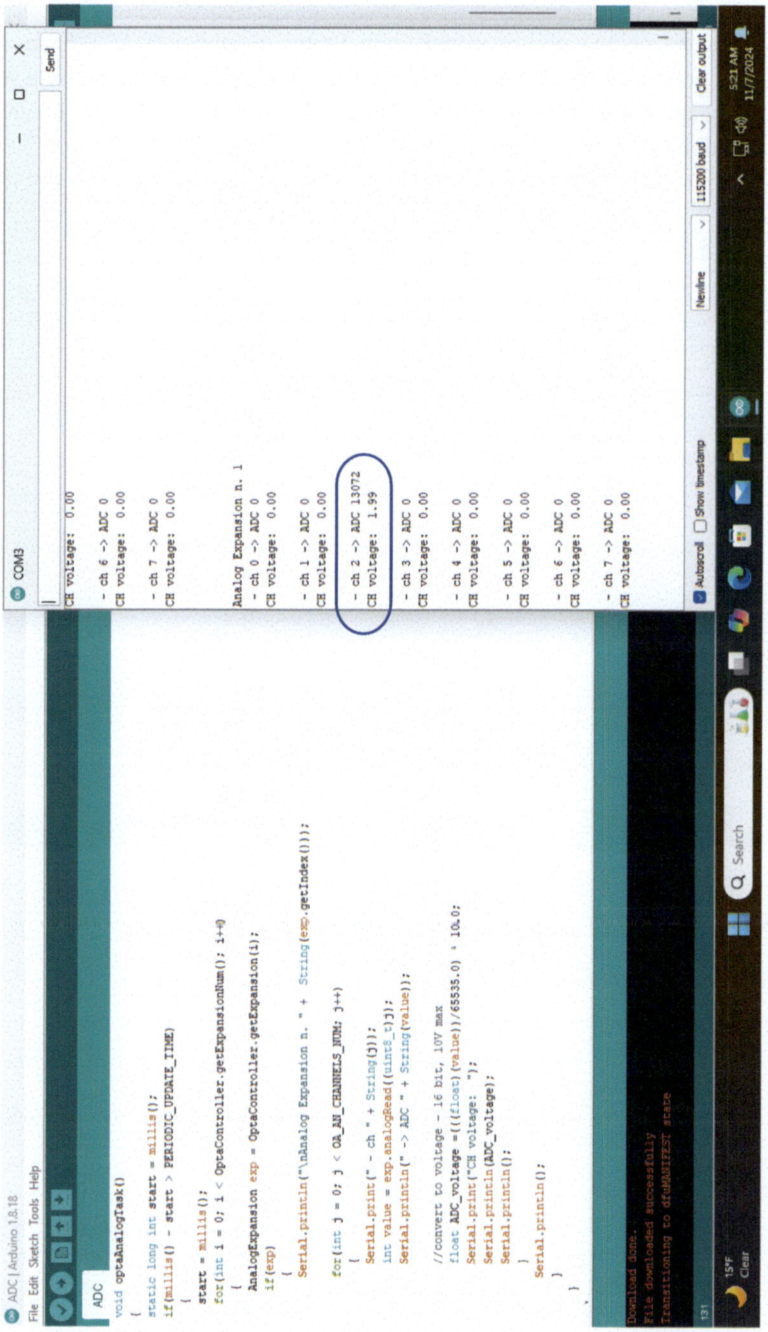

Fig. 6.8 Results of testing the ADC sketch (CC BY-NC-SA) (www.arduino.cc)

6.4.5 Digital–To–Analog (DAC)

The Opta Analog expansion is equipped with eight digital–to–analog (DAC) channels. These outputs may be configured as current or voltage DACs. The outputs are accessible via I1, I2, I3, I4, O1, I5, I6, and O2 terminals. The DAC outputs have 13 bits of resolution and have a voltage range from 0 to 11 V with a maximum current rating of 9 mA.

In the following sketch the DAC channels are configured as voltage DACs to provide a ramp voltage from 0 to 10 V. The resulting waveforms are shown in Fig. 6.9. The waveforms are captured using a DATAQ Instruments DI–1100 Data Acquisition Starter Kit (www.dataq.com).

```
//***************************************************************************
//FILE NAME:     simple DAC adapted from DAC.ino
//AUTHOR:        Daniele Aimo
//EMAIL:         d.aimo@arduino.cc
//DATE:          20231211
//DESCRIPTION:   This example shows how to use the OptaBlue library to control
//               the Opta Analog expansion. The example shows how to set the
//               DAC value for each channel. All channels are configured as
//               as voltage DACs.
//LICENSE:       Copyright (c) 2024 Arduino SA
//               This Source Code Form is subject to the terms of the Mozilla
//               Public License (MPL), v 2.0. You can obtain a copy of the MPL
//               at \url{http://mozilla.org/MPL/2.0/.}
//Modified:      S. Barrett, Nov 2024
//***************************************************************************

#include "OptaBlue.h"

#define PERIODIC_UPDATE_TIME 500
#define DELAY_AFTER_SETUP 200

void setup()
{
Serial.begin(115200);
delay(2000);
OptaController.begin();

for(int device = 0; device < OptaController.getExpansionNum(); device++)
  {
  for(int ch = 0; ch < OA_AN_CHANNELS_NUM; ch++)
    {
    AnalogExpansion::beginChannelAsDac(OptaController,
                                       device,
                                       ch,
                                       OA_VOLTAGE_DAC,
                                       true,
                                       false,
                                       OA_SLEW_RATE_0);
    }
  }
}


void loop()
{
OptaController.update();
printExpansionInfo();
```

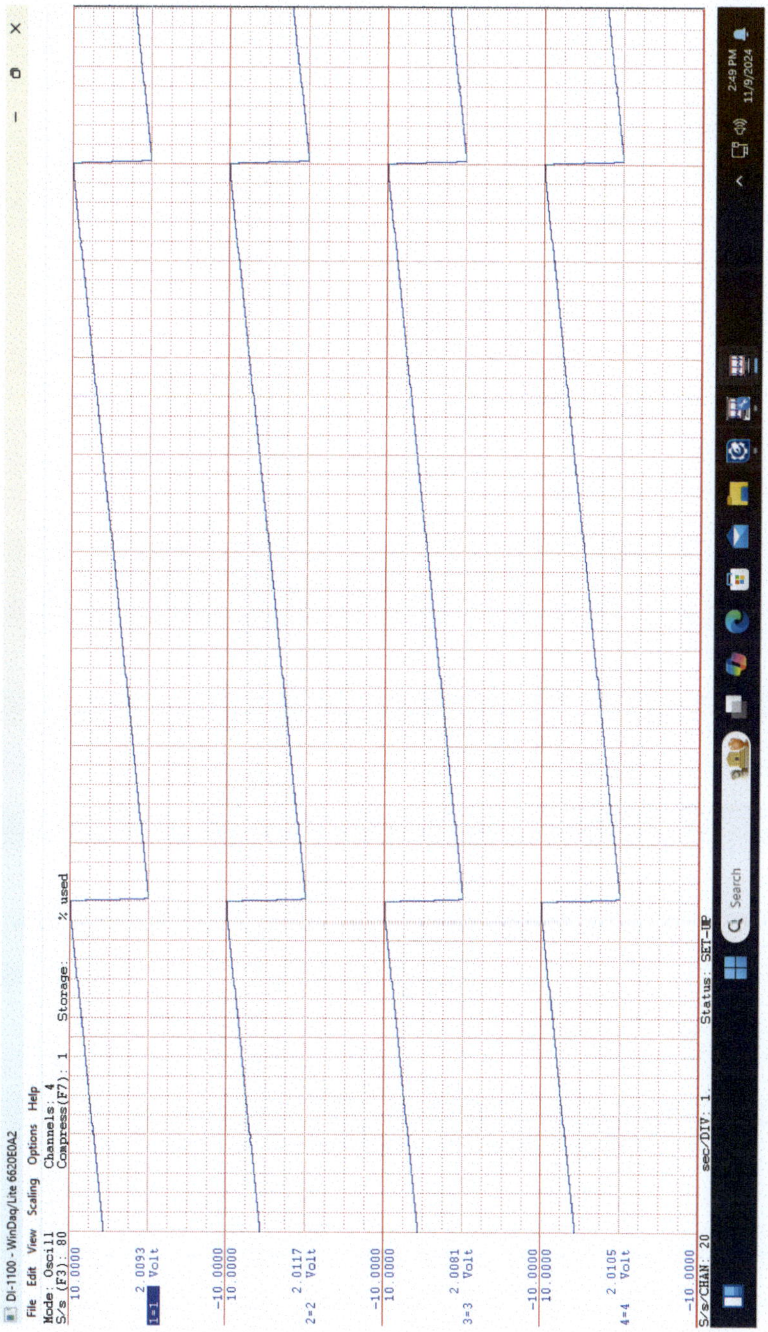

Fig. 6.9 Results of testing the DAC sketch (CC BY–NC–SA) (www.arduino.cc)

```
optaAnalogTask();
}

//*************************************************************************************

void printExpansionType(ExpansionType_t t)
{
if (t == EXPANSION_NOT_VALID)
  {
  Serial.print("Unknown!");
  }
else if (t == EXPANSION_OPTA_DIGITAL_MEC)
  {
  Serial.print("Opta --- DIGITAL [Mechanical]  ---");
  }
else if (t == EXPANSION_OPTA_DIGITAL_STS)
  {
  Serial.print("Opta --- DIGITAL [Solid State] ---");
  }
else if (t == EXPANSION_DIGITAL_INVALID)
  {
  Serial.print("Opta --- DIGITAL [!!Invalid!!] ---");
  }
else if (t == EXPANSION_OPTA_ANALOG)
  {
  Serial.print("~~~ Opta  ANALOG ~~~");
  }
else
  {
  Serial.print("Unknown!");
  }
}

//*************************************************************************************

void printExpansionInfo()
{
static long int start = millis();

if(millis() - start > 500)
  {
  start = millis();
  Serial.print("Number of expansions: ");
  Serial.println(OptaController.getExpansionNum());

  for(int i = 0; i < OptaController.getExpansionNum(); i++)
    {
    Serial.print("Expansion n. ");
    Serial.print(i);
    Serial.print(" type ");
    printExpansionType(OptaController.getExpansionType(i));
    Serial.print(" I2C address ");
     Serial.println(OptaController.getExpansionI2Caddress(i));
    }
  }
}

//*************************************************************************************

void optaAnalogTask()
{
static long int start = millis();
```

```
//using this the code inside the if will run every PERIODIC_UPDATE_TIME ms
//assuming the function is called repeatedly in the loop() function */

if(millis() - start > PERIODIC_UPDATE_TIME)
  {
  start = millis();
  static uint16_t dac_value = 0;
  dac_value += 100;                          //0.134 V step
    if(dac_value > 7500)                     //max value: 10.0 V
      {
      dac_value = 0;
      }

  for(int i = 0; i < OptaController.getExpansionNum(); i++)
    {
    AnalogExpansion exp = OptaController.getExpansion(i);
    if(exp)
      {
      Serial.println("Setting dac value " + String(dac_value) +
                     " on expansion n. " + String(exp.getIndex()));
      for(int ch = 0; ch < OA_AN_CHANNELS_NUM; ch++)
        {
         exp.setDac(ch, dac_value);
        }
      }
    }
  }
}

//*********************************************************************************
```

6.4.6 Resistor Temperature Detector (RTD) Temperature Measurement

There are several sensor types that may be used to measure temperature including integrated circuit (IC) based sensors, thermocouples, resistor temperature detectors (RTD), and thermistors.[3] In this example we use an RTD to measure ambient temperature.

Resistance Temperature Detectors or RTDs provide for the precise measurement of temperature. An RTD consists of a precision trimmed piece of metal or a coil of wire wrapped around a ceramic or glass core. The RTD is calibrated to have a specific resistance at a given temperature. For example, a PT100 RTD has a resistance of 100 Ω at 0 °C and a PT1000 has a resistance of 1000 Ω at 0 °C (Omega).

RTDs are available in a 2, 3, and 4–wire configuration. A 3–wire version is shown in Fig. 6.10a and b. Using a three or four wire configuration, the value of R_{RTD} resistance may be isolated from the resistance of the wire leads.

The connection of an RTD to the Opta analog expansion is shown in Fig. 6.10c. In the following sketch, a PT100 (Adafruit #3290) is used to measure temperature. The measured temperature is provided in Centigrade and Fahrenheit.

[3] We explore these sensors in "Arduino VIII: Machine Control".

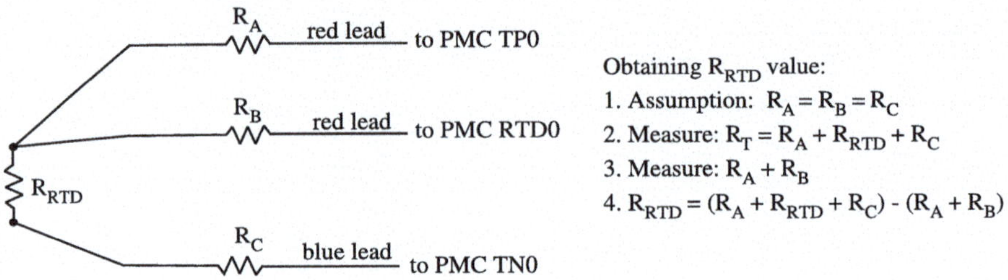

a) RTD 3-wire configuration.

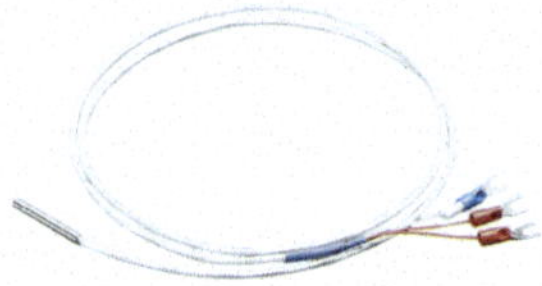

b) RTD 3-wire PT100.

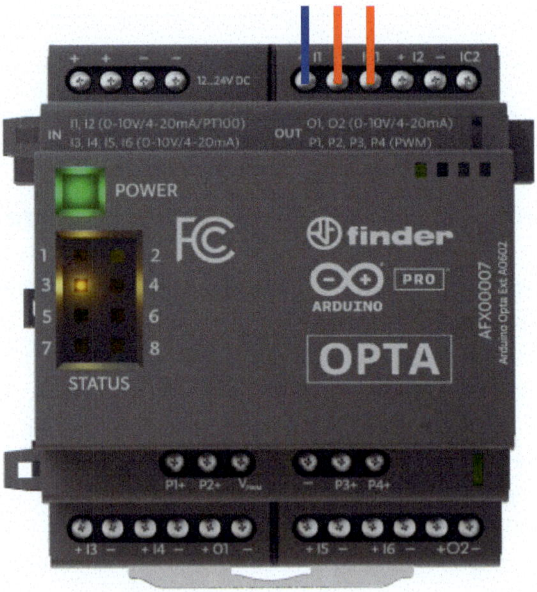

c) RTD connection to analog expansion.

Fig. 6.10 Resistance temperature detectors or RTDs

```
//*************************************************************************
//FILE NAME:    RTD.ino
//AUTHOR:       Daniele Aimo
//EMAIL:        d.aimo@arduino.cc
//DATE:         20231222
//DESCRIPTION: This example shows how to use RTD on Opta Analog expansion
//              driven by Opta OptaController.
//LICENSE:      Copyright (c) 2024 Arduino SA
//              This Source Code Form is subject to the terms fo the Mozilla
//              Public License (MPL), v 2.0. You can obtain a copy of the MPL
//              at \url{http://mozilla.org/MPL/2.0/.}
//NOTES:        In case of more than one Opta Analog expansion this sketch
//              works on the first one (the one closest to the OptaController)
//Modified:     S. Barrett, Nov 2024
//*************************************************************************

#include "OptaBlue.h"
#define PERIODIC_UPDATE_TIME 2000
#define DELAY_AFTER_SETUP 1000

// RTD constants
float a = 0.0039083;
float b = -0.0000005775;

int8_t oa_index = -1;

void setup()
{
Serial.begin(115200);
delay(2000);
Serial.println("*** Opta Analog RTD example ***");
OptaController.begin();

while (!Serial)                               //wait for serial monitor
  {
  ;
  }

for(int i = 0; i < OptaController.getExpansionNum(); i++)
  {
  for(int k = 0; k < OA_AN_CHANNELS_NUM;k++)
    {
    //all channels are initialized in the same way as RTD
    AnalogExpansion::beginChannelAsRtd(OptaController,
                            i, // the device
                            k, // the output channel you are using
                            true, // use 3 wire RTD
                            1.2); // current used on RTD in mA
    //set the sampling time for RTD to 1000 ms (default)
    AnalogExpansion::beginRtdUpdateTime(OptaController,
                            i,
                            1000);
    }
  }
}

void loop()
{
OptaController.update();
//printExpansionInfo();
optaAnalogTask();
}
```

```
//*****************************************************************************

void printExpansionType(ExpansionType_t t)
{
if(t == EXPANSION_NOT_VALID)
  {
  Serial.print("Unknown!");
  }
else if(t == EXPANSION_OPTA_DIGITAL_MEC)
  {
  Serial.print("Opta --- DIGITAL [Mechanical]  ---");
  }
else if(t == EXPANSION_OPTA_DIGITAL_STS)
  {
  Serial.print("Opta --- DIGITAL [Solid State] ---");
  }
else if(t == EXPANSION_DIGITAL_INVALID)
  {
  Serial.print("Opta --- DIGITAL [!!Invalid!!] ---");
  }
else if(t == EXPANSION_OPTA_ANALOG)
  {
  Serial.print("~~~ Opta  ANALOG ~~~");
  }
else
  {
  Serial.print("Unknown!");
  }
}

//*****************************************************************************

void printExpansionInfo()
{
static long int start = millis();

if(millis() - start > 5000)
  {
  start = millis();
  Serial.print("Number of expansions: ");
  Serial.println(OptaController.getExpansionNum());

  for(int i = 0; i < OptaController.getExpansionNum(); i++)
    {
    Serial.print("Expansion n. ");
    Serial.print(i);
    Serial.print(" type ");
    printExpansionType(OptaController.getExpansionType(i));
    Serial.print(" I2C address ");
    Serial.println(OptaController.getExpansionI2Caddress(i));
    }
  }
}

//*****************************************************************************

void optaAnalogTask()
{
static long int start = millis();
if(millis() - start > PERIODIC_UPDATE_TIME)
  {
  start = millis();
```

```
  for(int i = 0; i < OptaController.getExpansionNum(); i++)
    {
    AnalogExpansion aexp = OptaController.getExpansion(i);
    if(aexp)
      {
        Serial.println("Expansion n. " + String(aexp.getIndex()));
        for(int j = 0; j < 8; j++)
          {
          float value = aexp.getRtd((uint8_t)j);
          Serial.print("ch ");
          Serial.print(j);
          Serial.print(" -> ");
          Serial.print(value);
        Serial.print(" ?");
        float temp = (-(1.0 / 100.0) * (50.0 * a - 10*sqrt(b * value +
                        25.0 * pow(a, 2.0) - 100.0 * b))) / b;
        Serial.print(" -> ");
        Serial.print(temp);
        Serial.print(" C");
        Serial.println();
          Serial.print(" ");
          }
        Serial.println();
      }
    }
  }
}

//*****************************************************************************
```

6.4.7 Pulse Width Modulation (PWM)

The Opta analog expansion is equipped with four PWM channels designated P1 to P4. The PWM voltage is set by applying the desired voltage to the P_{PWM} terminal. The PWM duty cycle (0–100%) and baseline frequency (up to 10 kHz) is programmable. The maximum current output is 100 mA.

In the following sketch all four PWM channels are programmed with a varying duty cycle and hence a varying effective voltage. The circuit configuration and resulting waveforms are shown in Fig. 6.11. The waveforms are captured using a DATAQ Instruments DI-1100 Data Acquisition Starter Kit (www.dataq.com).

```
//*****************************************************************************
//FILE NAME:   Pwm.ino
//AUTHOR:      Daniele Aimo
//EMAIL:       d.aimo@arduino.cc
//DATE:        20231205
//DESCRIPTION: This sketch shows basic PWM usage with the OptaBlue library.
//LICENSE:     Copyright (c) 2024 Arduino SA
//             This Source Code Form is subject to the terms of the Mozilla
//             Public License (MPL), v 2.0. You can obtain a copy of the MPL
//             at \url{http://mozilla.org/MPL/2.0/.}
//Modified:    S. Barrett, Nov 2024
//*****************************************************************************

#include "OptaBlue.h"
```

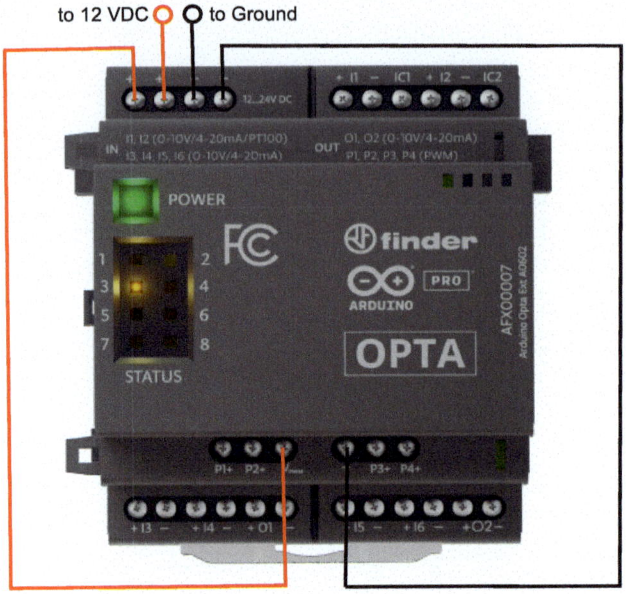

a) PWM power connection.

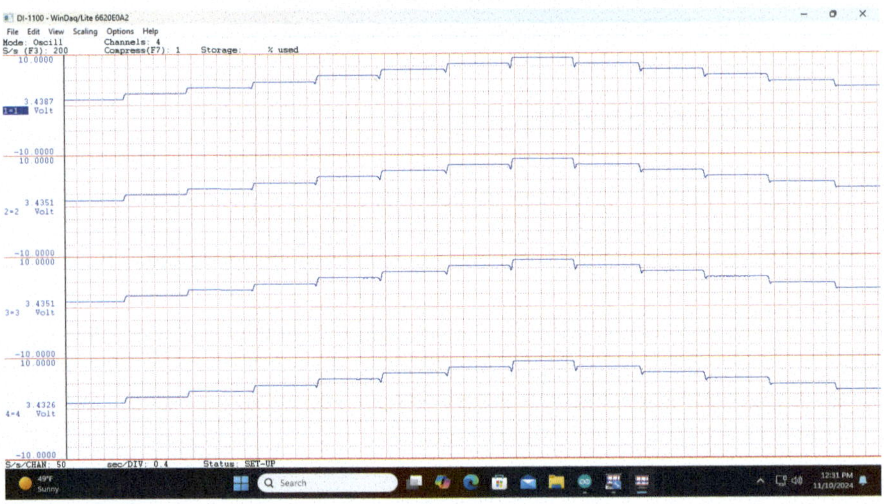

b) PWM result.

Fig. 6.11 Results of testing the PWM sketch (CC BY–NC–SA) (www.arduino.cc)

```
void setup()
{
Serial.begin(115200);
delay(2000);
OptaController.begin();
while (!Serial)                                          //wait for serial monitor
  {
  ;
  }
}

void loop()
{
OptaController.update();
printExpansionInfo();
optaAnalogTask();
}

//******************************************************************************

void printExpansionType(ExpansionType_t t)
{
if(t == EXPANSION_NOT_VALID)
  {
  Serial.print("Unknown!");
  }
else if(t == EXPANSION_OPTA_DIGITAL_MEC)
  {
  Serial.print("Opta --- DIGITAL [Mechanical]  ---");
  }
else if(t == EXPANSION_OPTA_DIGITAL_STS)
  {
  Serial.print("Opta --- DIGITAL [Solid State] ---");
  }
else if(t == EXPANSION_DIGITAL_INVALID)
  {
  Serial.print("Opta --- DIGITAL [!!Invalid!!] ---");
  }
else if(t == EXPANSION_OPTA_ANALOG)
  {
  Serial.print("~~~ Opta  ANALOG ~~~");
  }
else
  {
  Serial.print("Unknown!");
  }
}

//******************************************************************************

void printExpansionInfo()
{
static long int start = millis();

if(millis() - start > 5000)
  {
  start = millis();
  Serial.print("Number of expansions: ");
  Serial.println(OptaController.getExpansionNum());

  for(int i = 0; i < OptaController.getExpansionNum(); i++)
    {
```

```
    Serial.print("Expansion n. ");
    Serial.print(i);
    Serial.print(" type ");
    printExpansionType(OptaController.getExpansionType(i));
    Serial.print(" I2C address ");
    Serial.println(OptaController.getExpansionI2Caddress(i));
    }
  }
}

//*****************************************************************************

//The optaAnalogTask function runs every 2000 ms. It sets the pwm parameters
//for all the channels with a period equal to 10 ms (100 Hz) and a variable
//duty cycle from 10 to 70%

void optaAnalogTask()
{
static long int start = millis();
static bool stop_pwm = false;
if(millis() - start > 2000)
  {
  if(Serial.available())
    {
      while(Serial.available())
        {
        Serial.read();
        }
      stop_pwm = !stop_pwm;
    }
  start = millis();
  static uint16_t period = 10000;
  static uint16_t pulse = 0;
  static bool rising  = 1;
  if(rising)
    {
    pulse += 1000;
    if(pulse > 7000)
      {
        rising = 0;
      }
    }
  else
    {
    pulse -= 1000;
    if(pulse <= 1000)
      {
      rising = 1;
      }
    }

  for(int i = 0; i < OptaController.getExpansionNum(); i++)
    {
    AnalogExpansion aexp = OptaController.getExpansion(i);
    if(aexp)
      {
      if(stop_pwm)
        {
          Serial.println("PWM stopped");
          aexp.setPwm(OA_FIRST_PWM_CH, 0, pulse);
          aexp.setPwm(OA_FIRST_PWM_CH + 1, 0, pulse);
          aexp.setPwm(OA_FIRST_PWM_CH + 2, 0, pulse);
          aexp.setPwm(OA_FIRST_PWM_CH + 3, 0, pulse);
```

```
          }
        else
          {
          Serial.println("PWM started");
          aexp.setPwm(OA_FIRST_PWM_CH, period, pulse);
          aexp.setPwm(OA_FIRST_PWM_CH + 1, period, pulse);
          aexp.setPwm(OA_FIRST_PWM_CH + 2, period, pulse);
          aexp.setPwm(OA_FIRST_PWM_CH + 3, period, pulse);
          }
      }
    }
  for(int i = 0; i < OptaController.getExpansionNum(); i++)
    {
    AnalogExpansion aexp = OptaController.getExpansion(i);
    if(aexp)
      {
        Serial.println("PWM ch 0 period " +
        String(aexp.getPwmPeriod(OA_FIRST_PWM_CH))+
        " pulse " + aexp.getPwmPulse(OA_FIRST_PWM_CH));

        Serial.println("PWM ch 1 period " +
        String(aexp.getPwmPeriod(OA_FIRST_PWM_CH + 1)) +
        " pulse " + aexp.getPwmPulse(OA_FIRST_PWM_CH + 1));

        Serial.println("PWM ch 2 period " +
        String(aexp.getPwmPeriod(OA_FIRST_PWM_CH + 2)) +
        " pulse " + aexp.getPwmPulse(OA_FIRST_PWM_CH + 2));

        Serial.println("PWM ch 3 period " +
        String(aexp.getPwmPeriod(OA_FIRST_PWM_CH + 3))+
        " pulse " + aexp.getPwmPulse(OA_FIRST_PWM_CH + 2));
      }
    }
  }
}

//****************************************************************************
```

6.5 Application: Motor Speed Control with Pulse Width Modulation

As discussed in the Application section of Chap. 4, motor speed may be varied by changing the applied motor voltage. PWM control signal techniques may be used to precisely control the motor speed. With PWM the duty cycle and hence the effective voltage applied to the motor is varied.

In this example we revisit the Brother 12 VDC, 1,500 RPM DC motor equipped with an optical tachometer from Chap. 4. However, due to the 100 mA current limit of the PWM outputs (P1–P4), additional hardware interface electronics are required as shown in Fig. 6.12.

A software generated PWM signal is generated using the PWM sketch. The PWM signal is routed to the gate (G) of an IRF530 power MOSFET. The motor is connected between the 12 VDC motor power supply and the drain (D) connection. A reverse–biased 1N4001 diode is used for circuit protection. The MOSFET source (S) is connected to ground. Motor speed is monitored using techniques discussed in Chap. 4.

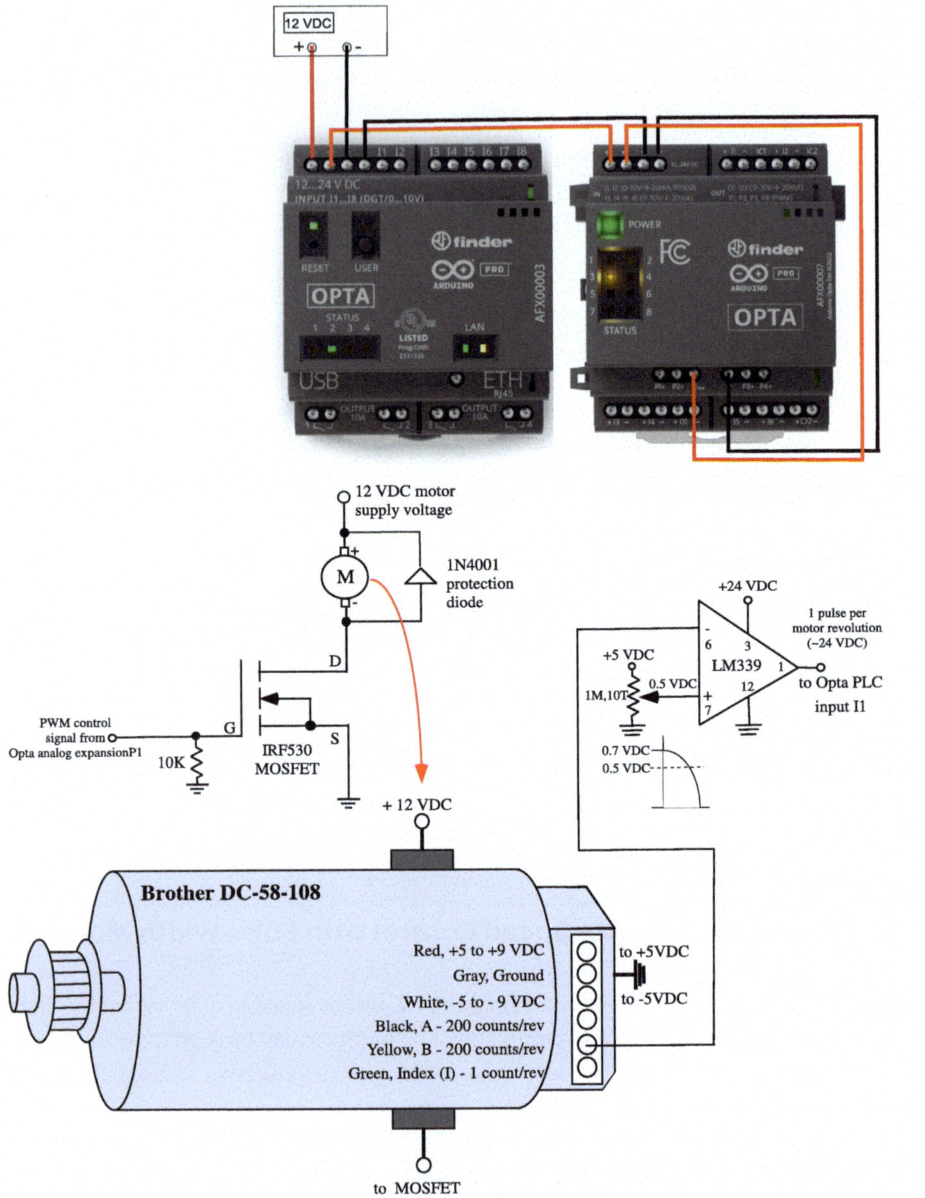

Fig. 6.12 DC motor demonstration circuit. Images used courtesy of the Arduino team (CC BY–NC––SA) (www.arduino.cc)

The goal of this example is to stabilize the speed of a DC motor using several different concepts discussed earlier in the book.
Activities:

1. Construct and test the interface circuit shown in Fig. 6.12.
2. Use the PWM sketch to vary the speed of the motor.
3. Write a test sketch to stabilize the DC motor at a specific motor speed. **Hint:** Adapt the motor speed control sketch provided in Chap. 4 and include PWM features from the Opta analog expansion.

6.6 Summary

In this chapter we provided a brief introduction to the recently released series of Opta PLC hardware expansion modules. The modules extend and enhance the Opta PLC capabilities.

6.7 Problems

1. Develop a one page summary of Opta digital and analog expansion modules. Include all related parameters (e.g. voltage, current, etc.).
2. Sketch a UML activity diagram for every sketch in this chapter.
3. Use the ADC sketch to measure ambient temperature using an LM34 Precision Fahrenheit Temperature Sensor manufactured by Texas Instruments. The LM34 provides 10 mV of output per degree Fahrenheit. Convert the LM34 output to temperature and display the result on the Serial Monitor in both Centigrade and Fahrenheit. Provide a circuit diagram and an Arduino sketch.
4. It is desired to monitor the output from the DAC sketch with 10 mm red LEDs ($V_f = 2V$, $I_f = 30$ mA). Provide a circuit diagram for the interface circuit and an Arduino sketch.
5. It is desired to monitor the output from the PWM sketch with 10 mm red LEDs ($V_f = 2V$, $I_f = 30$ mA). Provide a circuit diagram for the interface circuit and an Arduino sketch.
6. Within the Arduino Opta Blueprint Library is a sketch "genericDigital." Sketch a UML activity diagram for this sketch.
7. Within the Arduino Opta Blueprint Library is a sketch "genericAnalog." Sketch a UML activity diagram for this sketch.

References

1. D. Aimo, *Arduino Opta Blueprint Library,* ardunio.cc, 2024.
2. J. Bagur, C. Mendez, and J.C. Linares, *Opta User Manual*, ardunio.cc, 2024.
3. C. Mendez, *Opta Analog Expansion Usage with the PLC IDE*, ardunio.cc, 2024.
4. C. Mendez, *Opta Digital Expansions Usage with the PLC IDE*, ardunio.cc, 2024.
5. *RTD sensors–What are RTD temperature sensors?* www.omega.com , 2024.
6. *The I2C-Bus Specification*, Version 2.1, Philips Semiconductor, January 2000.

Safety

A

In this Appendix[1] a review of safety concepts are provided for ready reference. The information provided is condensed from the Electrical Safety Foundation International (ESFI) and several other sources. "The Electrical Safety Foundation International is the premier non–profit organization dedicated exclusively to promoting electrical safety at home and in the workplace. Since 1994, ESFI has led the way in promoting electrical safety across North America. Over the years, ESFI has become highly regarded by industry, media and consumer safety partners alike by constantly reinvigorating the way electrical safety is addressed. ESFI creates unique awareness and educational resources designed to meet the diverse needs of a variety of at–risk groups. The Electrical Safety Foundation International is dedicated exclusively to promoting electrical safety at home and in the workplace through education, awareness, and advocacy (https://esfi.org)."

In the next several sections we review the background information on the effects of electricity on the human body and present electrical safety procedures for safe equipment operation. You must strictly adhere to these procedures when working on book associated activities. With this in mind, we provide laboratory safe operating procedures and also electric shock treatment procedures.

Safety is paramount and is everybody's business!

[1] This Appendix was adapted with permission from "Arduino VI: Bioinstrumentation," S. Barrett, Springer, 2024.

S. F. Barrett, *Arduino VII*, Synthesis Lectures on Digital Circuits & Systems,
https://doi.org/10.1007/978-3-031-68609-2

A.1 Physiological Effects of Electricity

For an electrically induced physiological effect to occur, your body must become part of an electrical circuit. Current must enter at some point on your body and exit via another point. Through safe laboratory procedures and proper equipment design, current is prevented from entering the body.

From an electrical point of view the body is a resistor. The resistance of skin ranges from 15 KOhms to 1 MOhms per square centimeter. However, the resistance may decrease to one percent of its dry value when it is wet. Beyond the skin barrier, the internal resistance is approximately 200 Ω per limb and 100 Ω per trunk as shown in Fig. A.1. Application of Ohm's Law ($voltage = current \times resistance$) shows a voltage of 300 V applied across the body with dry skin will result in a current of approximately 10 mA. The same current flow will result at much lower values of voltage when your skin is moist or wet (Webster 2020).

When the body comes in contact with voltage, it is current flow through the body that causes physiological damage and potentially death. The body may be exposed to harmful currents via macroshock or microshock. In macroshock the body is exposed to a high voltage level. Microshocks provide low level, yet harmful currents to the heart. Microshocks result from leakage current within an AC powered circuit. Leakage currents are small currents that flow between two conductors at different potentials. Even if the conductors are well insulated, leakage currents may occur due to capacitive and resistive effects. Leakage currents flow into the body and the heart due to catheters or internal conductive monitoring leads (Webster 2020).

Fig. A.1 Body resistance

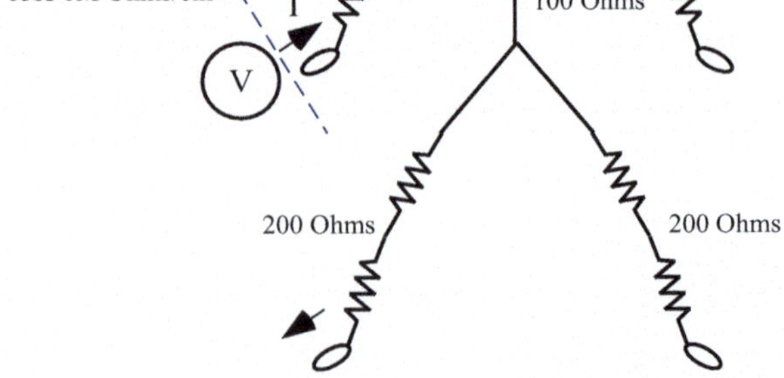

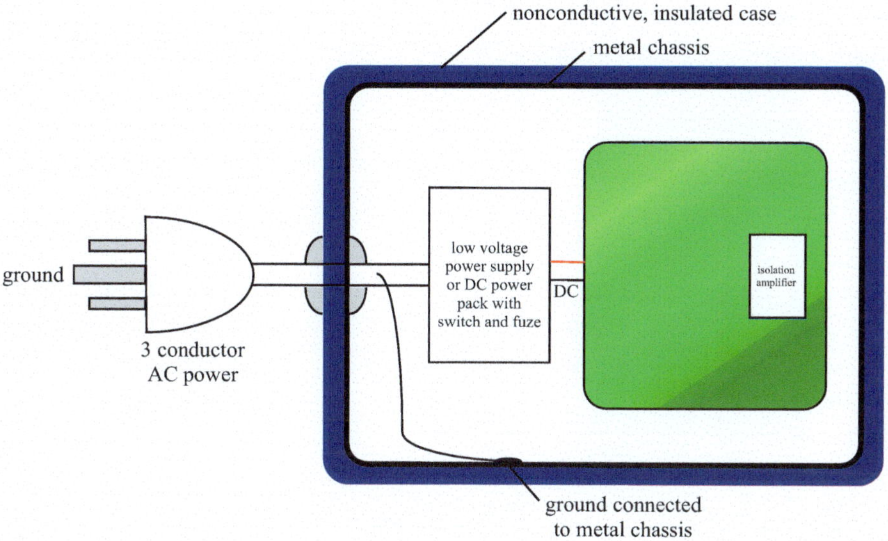

Fig. A.2 Safety design

Shock of either type is prevented via a combination of proper power system design and distribution, proper equipment grounding, proper equipment design, and safe operating practices and processes. As shown in Fig. A.2, a person is protected from both types of shock via these equipment design concepts (Webster 2020, Baretich 2015):

- A reliable equipment chassis ground to ground;
- Reduction of leakage current through properly insulated equipment conductors;
- Equipment double insulation with a nonconductive external covering or chassis;
- Equipment operation at the lower AC or DC voltage practical; and
- Use of isolation amplifiers when using conductive leads.

Equipment powered from an AC source must use a three conductor power cord. The ground conductor is connected to the equipment's metal chassis and also to Earth ground via a sound and reliable electrical distribution system. Also, equipment operation at the lowest possible AC or DC voltage further reduces shock hazard. The lower voltage coupled with a high, intact skin resistance reduce the potential from harmful current (Webster 2020).

Not much current flow is required to produce a physiological effect. The table below shows that currents as low as 0.5 mA illicit a perceptible physiological response. Note that as more current flows through the body the physiological effect is more severe (Webster 2020, ANSI/AAMI ESI–1993) (Table A.1).

Table A.1 60 Hz AC current response

Low range [mA]	High range [mA]	Physiological response
0.5	3	Start to feel the energy, tingling sensation
3	10	Experience pain, muscle contraction
10	40	Grip paralysis threshold–called "let–go" current
30	75	Respiratory systems shuts down
100	200	Experience heart fibrillation
200	500	Heart clamps tight
Over 1,500		Tissue and organs burn

"Let—go current" is the maximum current at which you can voluntarily withdraw from the current source. At higher current levels the muscles involuntarily contract and you can't let go of the current source. At current levels as low as 10 mA respiratory paralysis may result.

At approximately 50 mA ventricular fibrillation may result. Ventricular fibrillation is a condition where the ventricles of the heart quiver instead of pump. It is caused by an interruption of the normal electrical cardiac conduction cycle. Defibrillation action is usually required to bring the heart to a complete stop so that the normal cardiac conduction cycle may resume. At even higher levels of current sustained myocardial contraction and severe burns may occur. Currents as low as 10 mA through the body may result in death.

A.2 Electrical Safety Principles

The following is a list of electrical safety "do's" and "don'ts" compiled from several different electrical safety sources.

- Two or more individuals must be present in a laboratory at a given time to monitor one another's safety.
- Remove jewelry (rings, watches, necklaces, long earrings, etc.) when working with electrical or mechanical equipment.
- Watch out for loose wires.
- Work with one hand behind your back. This will prevent electrical current from entering one hand passing through your heart and out the other hand.
- Be aware of capacitors. The retain charge even when equipment is turned off.
- Use a three conductor electrical system with an Earth ground

- Keep fingers out of "live" chassis.
- Don't install components when the power is on.
- Don't work with wet or oily hands.
- Don't cut live wires.
- Double check circuit wiring before energizing a circuit
- Make sure that all members of your laboratory group realize when a circuit is being energized.
- Repair or replace leads with loose connectors.
- Avoid touching rotating parts or allowing wires to touch them.
- **Safety is everybody's business!**

A.3 Shock Rescue Procedures

In response to an electrical accident, follow these procedures immediately:

- Call for help.
- De—energize the circuit.
- Separate the person from the energy source.
- Make sure you and the victim are in a safe zone— not in contact with any electrical source, away from downed or broken wires.
- Never grab the person or pull the person off the current with your hands; you might become part of the circuit and become injured as well.
- Use a dry wood broom, leather belt, plastic rope or something similar that is non—conductive such as wood or plastic cane with hook on the end to free the person from the energy source.
- Administer first aid, apply mouth—to—mouth resuscitation and/or CPR; know what to do
- Keep the victim lying down, warm and comfortable to maintain body heat until help arrives. Do not move the person in case of injury to neck or back.
- If the victim is unconscious, put him/her on side to let fluids drain.
- Make sure the victim receives professional medical attention (person shocked could have heart failure hours later)

A.3.1 References

- ANSI/AAMI ESI—1993, *Safe Current Limits for Electromechanical Apparatus,* American National Standard, 1993.
- M. F. Baretich, *Electrical Safety Manual 2015—A Comprehensive Guide to Electrical Safety Standards for Healthcare Facilities,* American Safety in Medical Technology (AAMI), https://www.aami.org.
- *Electrical Safety Foundation International (ESFI),* https://esfi.org.
- J. G. Webster and A. J. Nimunkar, *Medical Instrumentation Application and Design,* fifth edition, John Wiley and Sons, Inc, 2020.

Embedded Systems Design

B

Objectives: After reading this appendix, the reader should be able to do the following:

- Define an embedded system;
- List all aspects related to the design of an embedded system;
- Provide a step–by–step approach to embedded system design;
- Discuss design tools and practices related to embedded systems design; and
- Apply embedded system design practices in the design of a microcontroller system employing several interacting subsystems.

B.1 Overview

In this chapter,[2] we begin with a definition of just what is an embedded system. We then explore the process of how to successfully (and with low stress) develop an embedded system prototype that meets established requirements.

[2] The information on embedded system design first appeared in "Microcontroller Fundamentals for Engineers and Scientists," Morgan and Claypool Publishers, 2006. It has been adapted with permission. Although first developed for embedded systems design, concepts provided here apply to Opta—based PLC system design.

S. F. Barrett, *Arduino VII*, Synthesis Lectures on Digital Circuits & Systems,
https://doi.org/10.1007/978-3-031-68609-2

B.2 What is an Embedded System?

An embedded system contains a microcontroller to accomplish its job of processing system inputs and generating system outputs. The link between system inputs and outputs is provided by a coded algorithm stored within the processor's resident memory. What makes embedded systems design so interesting and challenging is the design must also take into account the proper electrical interface for the input and output devices, limited on—chip resources, human interface concepts, the operating environment of the system, cost analysis, related standards, and manufacturing aspects (Anderson). Through careful application of this material you will be able to design and prototype embedded systems.

B.3 Embedded System Design Process

In this section, we provide a step—by—step approach to develop the first prototype of an embedded system that will meet established requirements. There are many formal design processes that we could study. We concentrate on the steps that are common to most. We purposefully avoid formal terminology of a specific approach and instead concentrate on the activities that are accomplished as a system prototype is developed. The design process we describe is illustrated in Fig. B.1 using a Unified Modeling Language (UML) activity diagram. We discuss the UML activity diagrams later in the appendix.

B.3.1 Project Description

The goal of the project description step is to determine what the system is ultimately supposed to do. To achieve this step you must thoroughly investigate what the system is supposed to do. Questions to raise and answer during this step include but are not limited to the following:

- What is the system supposed to do?
- Where will it be operating and under what conditions?
- Are there any restrictions placed on the system design?

To answer these questions, the designer interacts with the client to ensure clear agreement on what is to be done. If you are completing this project for yourself, you must still carefully and thoughtfully complete this step. The establishment of clear, definable system requirements may require considerable interaction between the designer and the client. It is essential that both parties agree on system requirements before proceeding further in the design process. The final result of this step is a detailed listing of system requirements and related specifications.

Fig. B.1 Embedded system design process

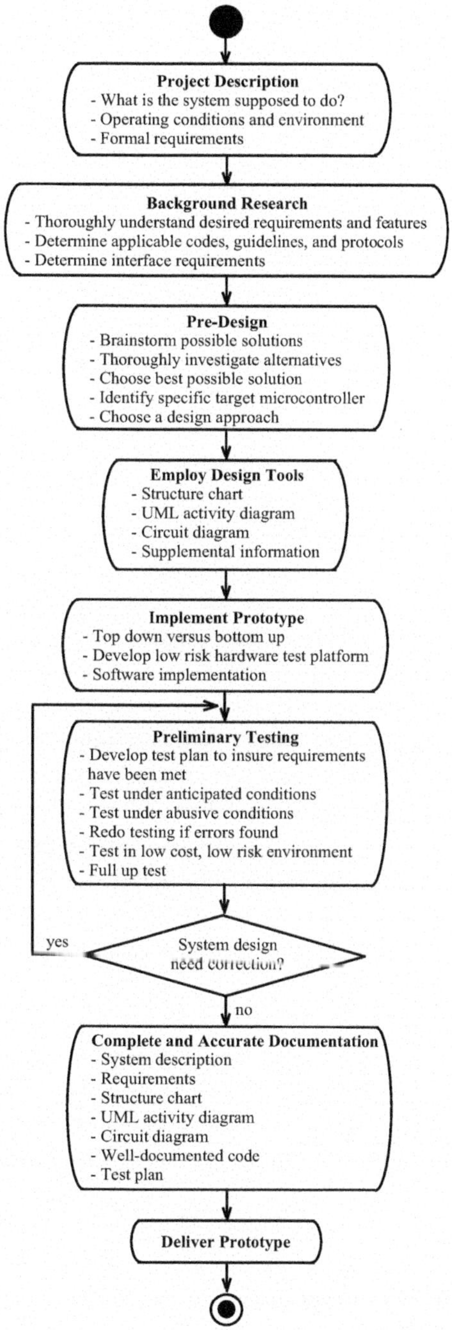

B.3.2 Background Research

Once a detailed list of requirements has been established, the next step is to perform background research related to the design. In this step, the designer will ensure they understand all requirements and features required by the project. This will again involve interaction between the designer and the client. The designer will also investigate applicable codes, guidelines, protocols, and standards related to the project. This is also a good time to start thinking about the interface between different portions of the project particularly the input and output devices peripherally connected to the microcontroller. The ultimate objective of this step is to have a thorough understanding of the project requirements, related project aspects, and any interface challenges within the project.

B.3.3 Pre—Design

The goal of the pre—design step is to convert a thorough understanding of the project into possible design alternatives. Brainstorming is an effective tool in this step. Here, a list of alternatives is developed. Since an embedded system typically involves both hardware and/or software, the designer can investigate whether requirements could be met with a hardware only solution or some combination of hardware and software. Generally, speaking a hardware only solution executes faster; however, the design is somewhat fixed once fielded. On the other hand, a software implementation provides flexibility and a typically slower execution speed. Most embedded design solutions will use a combination of both hardware and software to capitalize on the inherent advantages of each.

Once a design alternative has been selected, the general partition between hardware and software can be determined. It is also an appropriate time to select a specific hardware device to implement the prototype design. If a microcontroller technology has been chosen, it is now time to select a specific controller. This is accomplished by answering the following questions:

- What microcontroller systems or features i.e., ADC, PWM, timer, etc.) are required by the design?
- How many input and output pins are required by the design?
- What is the maximum anticipated operating speed of the microcontroller expected to be?

B.3.4 Design

With a clear view of system requirements and features, a general partition determined between hardware and software, and a specific microcontroller chosen, it is now time to tackle the actual design. It is important to follow a systematic and disciplined approach to design. This will allow for low stress development of a documented design solution that meets requirements. In the design step, several tools are employed to ease the design process. They include the following:

- Employing a top—down design, bottom up implementation approach,
- Using a structure chart to assist in partitioning the system,
- Using a Unified Modeling Language (UML) activity diagram to work out program flow, and
- Developing a detailed circuit diagram of the entire system.

Let's take a closer look at each of these. The information provided here is an abbreviated version of the one provided in "Microcontrollers Fundamentals for Engineers and Scientists." The interested reader is referred there for additional details and an in—depth example (Barrett and Pack).

Top down design, bottom up implementation. An effective tool to start partitioning the design is based on the techniques of top—down design, bottom—up implementation. In this approach, you start with the overall system and begin to partition it into subsystems. At this point of the design, you are not concerned with how the design will be accomplished but how the different pieces of the project will fit together. A handy tool to use at this design stage is the structure chart. The structure chart shows the hierarchy of how system hardware and software components will interact and interface with one another. You should continue partitioning system activity until each subsystem in the structure chart has a single definable function.

UML Activity Diagram. Once the system has been partitioned into pieces, the next step in the design process is to start working out the details of the operation of each subsystem we previously identified. Rather than beginning to code each subsystem as a function, we will work out the information and control flow of each subsystem using another design tool: the Unified Modeling Language (UML) activity diagram. The activity diagram is simply a UML compliant flow chart. UML is a standardized method of documenting systems. The activity diagram is one of the many tools available from UML to document system design and operation. The basic symbols used in a UML activity diagram for a microcontroller based system are provided in Fig. B.2 (Fowler).

To develop the UML activity diagram for the system, we can use a top—down, bottom—up, or a hybrid approach. In the top—down approach, we begin by modeling the overall flow of the algorithm from a high level. If we choose to use the bottom—up approach, we would begin at the bottom of the structure chart and choose a subsystem for flow modeling. The specific course of action chosen depends on project specifics. Often, a combination of both techniques, a hybrid approach, is used. You should work out all algorithm details at the UML activity diagram level prior to coding any software. If you can not explain system operation at this higher level, first, you have no business being down in the detail of devel-

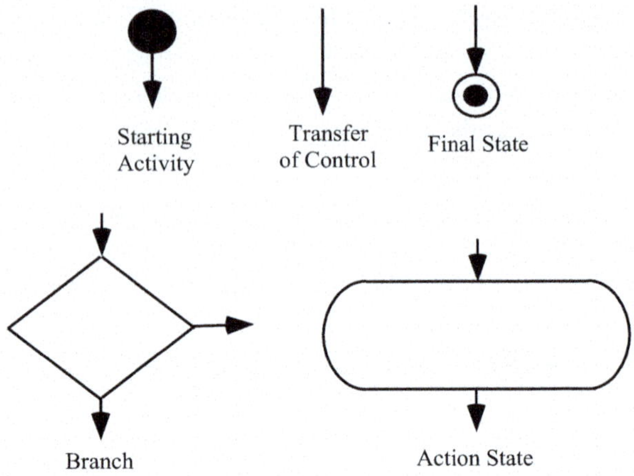

Fig. B.2 UML activity diagram symbols. Adapted from (Barrett and Pack)

oping the code. Therefore, the UML activity diagram should be of sufficient detail so you can code the algorithm directly from it (Dale).

In the design step, a detailed circuit diagram of the entire system is developed. It will serve as a roadmap to implement the system. It is also a good idea at this point to investigate available design information relative to the project. This would include hardware design examples, software code examples, and application notes available from manufacturers. At the completion of this step, the prototype design is ready for implementation and testing.

B.3.5 Implement Prototype

To successfully implement a prototype, an incremental approach should be followed. Again, the top—down design, bottom—up implementation provides a solid guide for system implementation. In an embedded system design involving both hardware and software, the hardware system including the microcontroller should be assembled first. This provides the software the required signals to interact with. As the hardware prototype is assembled on a prototype board, each component is tested for proper operation as it is brought online. This allows the designer to pinpoint malfunctions as they occur.

Once the hardware prototype is assembled, coding may commence. As before, software should be incrementally brought online. You may use a top down, bottom up, or hybrid approach depending on the nature of the software. The important point is to bring the software online incrementally such that issues can be identified and corrected early on.

It is highly recommended that low cost stand—in components be used when testing the software with the hardware components. For example, push buttons, potentiometers, and LEDs may be used as low cost stand—in component simulators for expensive input instrumentation devices and expensive output devices such as motors. This allows you to insure the software is properly operating before using it to control the actual components.

B.3.6 Preliminary Testing

To test the system, a detailed test plan must be developed. Tests should be developed to verify that the system meets all of its requirements and also intended system performance in an operational environment. The test plan should also include scenarios in which the system is used in an unintended manner. As before a top—down, bottom—up, or hybrid approach can be used to test the system.

Once the test plan is completed, actual testing may commence. The results of each test should be carefully documented. As you go through the test plan, you will probably uncover a number of run time errors in your algorithm. After you correct a run time error, the entire test plan must be performed again. This ensures that the new fix does not have an unintended effect on another part of the system. Also, as you process through the test plan, you will probably think of other tests that were not included in the original test document. These tests should be added to the test plan. As you go through testing, realize your final system is only as good as the test plan that supports it!

Once testing is complete, you might try another level of testing where you intentionally try to "jam up" the system. In another words, try to get your system to fail by trying combinations of inputs that were not part of the original design. A robust system should continue to operate correctly in this type of an abusive environment. It is imperative that you design robustness into your system. When testing on a low cost simulator is complete, the entire test plan should be performed again with the actual system hardware. Once this is completed you should have a system that meets its requirements!

B.3.7 Complete and Accurate Documentation

With testing complete, the system design should be thoroughly documented. Much of the documentation will have already been accomplished during system development. Documentation will include the system description, system requirements, the structure chart, the UML activity diagrams documenting program flow, the test plan, results of the test plan, system schematics, and properly documented code. To properly document code, you should carefully comment all functions describing their operation, inputs, and outputs. Also, comments should be included within the body of the function describing key portions of the code. Enough detail should be provided such that code operation is obvious. It is also extremely helpful to provide variables and functions within your code names that describe their intended use.

You might think that a comprehensive system documentation is not worth the time or effort to complete it. Complete documentation pays rich dividends when it is time to modify, repair, or update an existing system. Also, well—documented code may be often reused in other projects: a method for efficient and timely development of new systems.

B.4 Summary

In this appendix, we discussed the design process, related tools, and applied the process to a real world design. It is essential to follow a systematic, disciplined approach to embedded systems design to successfully develop a prototype that meets established requirements.

B.5 References

- M. Anderson, Help Wanted: Embedded Engineers Why the United States is losing its edge in embedded systems, *IEEE–USA Today's Engineer*, Feb 2008.
- S. Barrett, D. Pack, *Microcontrollers Fundamentals for Engineers and Scientists* (Morgan and Claypool Publishers, 2006). 10.2200/S00025ED1V01Y200605DCS001.
- M. Fowler with K. Scott "UML Distilled – A Brief Guide to the Standard Object Modeling Language," 2nd edition. Boston:Addison–Wesley, 2000.
- N. Dale and S.C. Lilly "Pascal Plus Data Structures," 4th edition. Englewood Cliffs, NJ: Jones and Bartlett, 1995.

Index

S. F. Barrett, *Arduino VII*, Synthesis Lectures on Digital Circuits & Systems, https://doi.org/10.1007/978-3-031-68609-2

The manufacturer's authorised representative in the EU is Springer Nature Customer Service Centre GmbH, Europaplatz 3, 69115 Heidelberg, Germany. If you have any concerns regarding our products, please contact ProductSafety@springernature.com

Printed and bound by CPI Group (UK) Ltd, Croydon, CR0 4YY
30/09/2026
02225233-0018